T0303298

Air Force Manpower Requirements and Component Mix

A Focus on Agile Combat Support

Albert A. Robbert, Lisa M. Harrington, Tara L. Terry, Hugh G. Massey

RAND Project AIR FORCE

Prepared for the United States Air Force
Approved for public release; distribution unlimited

RAND CORPORATION

For more information on this publication, visit www.rand.org/t/RR617

Library of Congress Cataloging-in-Publication Data is available for this publication.

ISBN: 978-0-8330-8559-7

Published by the RAND Corporation, Santa Monica, Calif.

Preface

Processes for determining U.S. Air Force manpower requirements vary considerably across and within the variety of workforces employed to meet Air Force missions. The relevant workforces include active duty military personnel, full-time and part-time Air Force Reserve (AFR) and Air National Guard (ANG) military personnel, civilian employees, and contractors. Distinctive processes have been developed for quantifying needs for operational, maintenance, and non-maintenance agile combat support workforces. Prioritization among requirements is reached through the programming and budgeting processes of the Air Force Corporate Structure, but that structure depends on the development of manpower standards, manpower modeling, and other similar processes to identify the unconstrained manpower costs of executing and supporting various missions. Our primary focus in this report is on those quantitatively oriented manpower requirements processes and the extent to which they are validated, coordinated, and consistent. We also explore the qualitative side of personnel requirements, whereby, through specialty classification standards and other related processes, the required competencies of various workforces are specified.

The research reported here was commissioned by the Deputy Chief of Staff for Manpower, Personnel, and Services, United States Air Force, and conducted within the Manpower, Personnel, and Training Program of RAND Project AIR FORCE as part of a fiscal year 2013 project titled "Personnel Requirements and Component Mix."

RAND Project AIR FORCE

RAND Project AIR FORCE (PAF), a division of the RAND Corporation, is the U.S. Air Force's federally funded research and development center for studies and analyses. PAF provides the Air Force with independent analyses of policy alternatives affecting the development, employment, combat readiness, and support of current and future air, space, and cyber forces. Research is conducted in four programs: Force Modernization and Employment; Manpower, Personnel, and Training; Resource Management; and Strategy and Doctrine. The research reported here was prepared under contract FA7014-06-C-0001.

Additional information about PAF is available on our website:
http://www.rand.org/paf/

Contents

Figures

Tables

Summary

In the research underlying this document, we explored policies regarding manpower and personnel requirements processes and sought to understand their robustness. Since some requirements are based on wartime or deployment needs rather than peacetime or garrison needs, we sought to determine if a common-sight picture of wartime demands was available.

We found that manpower processes differ across components (active, Air National Guard [ANG], and Air Force Reserve [AFR]), functional areas (operations, maintenance, and combat support), and workforces (active military, active Guard/Reserve [AGR], air reserve technician [ART], dual-status technician [ANG equivalent of the AFR's ART], traditional part-time reservist/guardsman [TR], individual mobilization augmentee [IMA], civilian, and contractor). The standard processes that address requirements in various components, functional areas, and workforces include manpower standards, crew ratios, the Logistics Composite Model (LCOM), technical estimates, deployment requirements expressed in unit type codes (UTCs), headquarters staff strength ceilings, procedures for IMA requirements, in-house versus contractor competitions conducted under Office of Management and Budget (OMB) Circular A-76, and other minor processes.

Within this collection of standard processes, we judged four processes (manpower standards, crew ratios, LCOM, and technical estimates) to be more rigorous than the others. These processes feature formulas or models whose inputs include workload drivers and workload-to-manpower relationships that have been derived empirically or through expert judgment, yielding objective, replicable results. Figure S.1 indicates the coverage of manpower authorizations by rigorous or standard requirements processes in various workforces. The UTC process for determining part-time reserve-component requirements would be considered rigorous if it were consistently applied. As discussed below, however, application is inconsistent. Accordingly, we judged authorizations based on UTCs to be *potentially* rigorous (identified by crosshatched bar segments in Figure S.1).

The limited coverage of authorizations by rigorous or other standard processes is attributable, in some cases, to the unique nature or limited scope of some missions. Where limited manpower resources are available, priority must be given to developing requirements processes for larger, more prevalent missions. We note, however, that the resources of the Air Force's manpower requirements squadrons and flights appear to be inadequate to their task, as evidenced by both the limited coverage of requirements by standard processes and the age distribution of current manpower standards. Some standards are old enough to raise questions regarding their continued validity.

Figure S.1. Rigorous and Standard Process Coverage by Workforce Categories (September 2012)

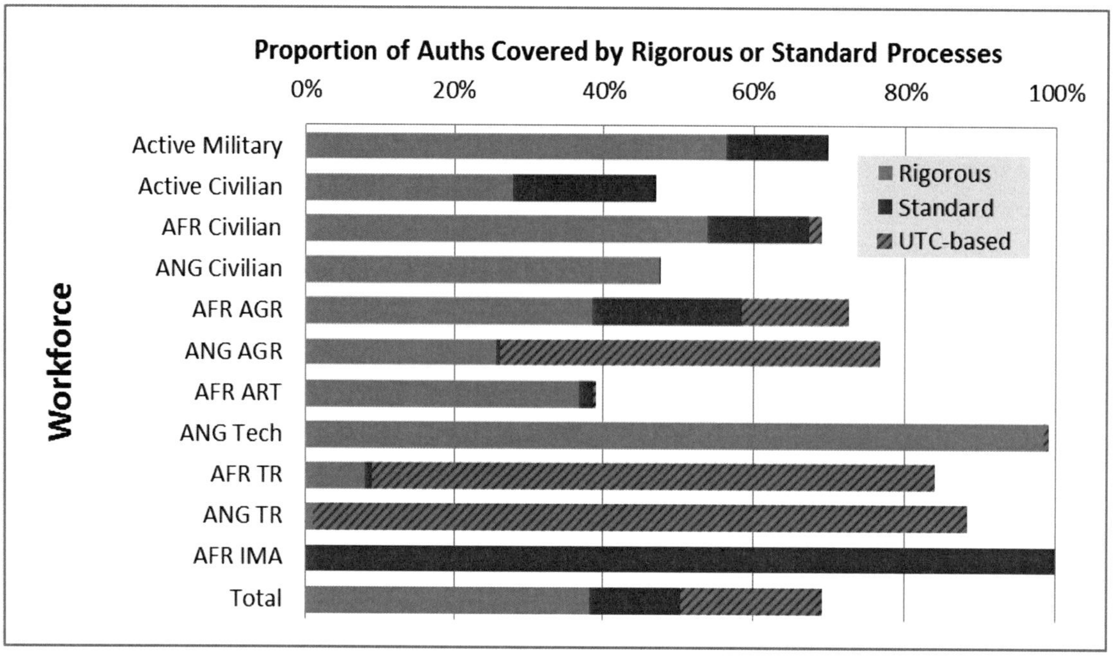

SOURCE: Manpower Programming and Execution System (MPES).
NOTES: Includes funded authorizations only and excludes permanent party student authorizations. ANG Tech = ANG dual-status technician. Workforce sizes were as follows:

Active Military	301,307	AFR AGR	2,938	AFR TR	46,817
Active Civilian	149,934	ANG AGR	15,007	ANG TR	91,447
AFR Civilian	3,795	AFR ART	10,584	AFR IMA	8,711
ANG Civilian	1,588	ANG Tech	23,165	Total	655,293

Another area of concern is the separation of manpower standards by component. Where active-component (AC) and reserve-component (RC) units operate independently, this is less of an issue. However, when AC and RC units are located on the same installation, and even more so when they are associated organizationally in an equipment-sharing arrangement, we would expect some efficiencies to be found in developing consolidated manpower standards for AC and full-time RC workforces.

Requirements for part-time, unit-assigned, drilling RC workforces, referred to as *traditional reservists/guardsmen*, are based primarily on wartime requirements. We found, however, that Air Force processes for determining wartime requirements for non-maintenance agile combat support (ACS) functions have lost rigor.[1] Efforts are underway to revitalize parts of the process, with primary emphasis on a wartime workforce planning model, Strategic Manpower Assessment of Requirements–Total Force (SMARTForce), that relies on functional manager

---

[1] ACS is the term used by the Air Force to denote non-operational support functions. In some contexts it includes aircraft and missile maintenance functions and in others it does not. In this document, we exclude these maintenance workforces from our definition of ACS.

inputs in lieu of the more customary time-phased force and deployment documents (TPFDDs) to estimate the UTCs required for various planning scenarios. As wartime requirements become clearer, and as our analysis of the past decade of relatively intense deployment demands indicates, TR requirements in most ACS functions appear to be in excess of need. Considerable savings are possible, but they cannot be confirmed without a firmer set of wartime requirements.

Using the RC as an operational reserve has important implications for the mix of AC and RC manpower, but the manpower cost on the RC side is higher than necessary due to restrictions on the duties of RC personnel paid from Reserve Personnel Appropriation (RPA) funding. The restrictions generally limit RPA-funded activities to maintaining equipment or organizing, administering, recruiting, instructing, and training reserve forces (the so-called OARIT duties). These restrictions tend to mandate more training than is needed in many circumstances and invite circumventions to allow greater participation by reservists in active missions.

A related concern is the tenuous linkage of IMA authorizations to wartime requirements. Each IMA position is individually justified through a process that provides no linkages to UTCs, which are the basic building blocks of wartime requirements. Additionally, the distribution of IMA requirements across officer/enlisted and career-field categories bears no discernible relationship to historic wartime deployment demands.

In addition to examining quantitative requirements processes, we documented qualitative processes. The primary vehicle for qualitative requirements is the Air Force classification system, wherein the knowledge, skills, abilities, training, and education required for various specialties are specified. Other qualitative requirements, including advanced education, language, special experience, and security clearance, can be found in individual manpower position records contained in MPES. Among these requirements, we saw the need for additional attention to officer education prerequisites.

Functional managers and career field managers play a pivotal role in many manpower processes. To gain their perspective, we interviewed representatives of three functional areas: civil engineering; security forces; and manpower, personnel, and services. A common issue raised in these interviews was the lack of a standard Air Force process for determining wartime requirements, which compels each career field manager to make an independent assessment of those requirements. We found that the assumptions used by these three functional area managers differed in important ways, yielding inconsistent assessments of wartime requirements. We also found little or no coordination between AC and RC functional area managers regarding the resourcing of wartime UTC requirements.

Our recommendations touch on broader manpower requirements processes, but focus most heavily on AC/RC interfaces and the determination of wartime requirements.

- Improve existing manpower requirements processes. Consider additional resources and/or the elevation of competency requirements in the workforces of the Air Force's manpower requirements squadrons.

- Refine the policies and practices for sizing wartime requirements. A revitalized, deliberate planning process would provide a stronger basis for wartime-related manpower requirements.
- Permit more flexibility in the use of RC resources. Legislative changes are needed to allow more latitude in using RC workforces to make operational contributions to active missions.
- Refine the approaches for determining TR requirements. Greater visibility of UTC requirements and more disciplined AC/RC sourcing of UTCs can significantly reduce total force manpower costs.
- Strengthen the linkages between IMA authorizations and wartime and other requirements. IMA requirements not demonstrably linked to expected needs should be eliminated.
- Improve the specification of officer education requirements. More specific undergraduate degree requirements are needed to guide officer accessions; a shift in the grade distribution of advanced degree requirements is needed to better utilize available degree holders.

Acknowledgments

The research underlying this report was launched by Mr. Bob Corsi, the Air Force Assistant Deputy Chief of Staff for Manpower, Personnel, and Services, whose insight into Air Force manpower processes indicated to him that a comprehensive review of manpower requirements processes was overdue. The need for a review became more compelling when the Secretary of the Air Force asked for an assessment of personnel readiness, which depends on a sound understanding of wartime requirements. Ms. Michelle LoweSolis, Director of Plans and Integration (AF/A1X), assumed primary responsibility for the personnel readiness review and as the sponsor for our research, leading us to work closely with members of her staff, principally Colonel Dennis Curran and Mr. Alex Hodge.

The Directorate of Manpower, Organization, and Resources (AF/A1M) also had important equities in the research. Its director and deputy director, Brigadier General Richard Murphy and Mr. Bill Snodgrass, and action officers, Mr. Greg Parton and Mr. Gene Blinn, were heavily engaged with us throughout the project. Mr. Darryl Hickman, Director of Resources Analyses (AF/A9R), and others in the Air Force Directorate of Analyses, Assessments, and Lessons Learned (AF/A9) also contributed immeasurably to our efforts to understand wartime requirements processes.

Special thanks go to Mr. Greg Cummings, Air Force Civil Engineer Center/CCX, Mr. Kevin Allen from the Directorate of Security Forces (AF/A7SX), and, from the Directorate of Plans and Integration (AF/A1XR), Colonel Chris Oleksa and Mr. Rod Ballard, who serve as functional managers in the civil engineering, security forces, and manpower/personnel/services communities, for taking time to explain their processes and share their perspectives with us.

Our thoughts and words benefitted from careful reviews by Mr. Patrick Mills, Dr. Dave Graham, and Mr. Al Schroetel and editing by Melissa McNulty.

Abbreviations

AC	active component
ACS	agile combat support
AEF	air expeditionary force
AF/A1	Air Force Deputy Chief of Staff for Manpower, Personnel, and Services
AF/A1M	Air Force Directorate of Manpower, Organization, and Resources
AF/A1MR	Air Force Manpower Requirements Division
AF/A5XW	Air Force War Planning and Policy Division
AF/A8XF	Air Force Total Force Management Division
AF/A9	Air Force Directorate of Analyses, Assessments, and Lessons Learned
AFB	Air Force Board
AFC	Air Force Council
AFECD	Air Force Enlisted Classification Directory
AFG	Air Force Group
AFI	Air Force Instruction
AFJQMS	Air Force job qualification standard
AFMAN	Air Force Manual
AFOCD	Air Force Officer Classification Directory
AFPC/MA	Air Force Personnel Center Directorate of Manpower
AFR	Air Force Reserve
AFRC	Air Force Reserve Command
AFRC/A1M	AFRC Manpower, Organization, and Resources Division
AFRCI	AFRC Instruction
AFSC	Air Force specialty code
AGR	Active Guard/Reserve
ANG	Air National Guard
ANG/A1M	ANG Manpower, Organization, and Resources Division
ANGI	ANG Instruction
ART	air reserve technician
CE	civil engineering
CFETP	career field education and training plan
CME	contractor manpower equivalent

COCOM	combatant commander
CONUS	continental United States
DAF	Department of the Air Force
DAV	deployment availability code
DMDC	Defense Manpower Data Center
DoD	Department of Defense
DoDI	Department of Defense Instruction
DPG	Defense Planning Guidance
FAM	functional area manager
FES	fire emergency service
FORSIZE/BLA	Force Sizing Exercise/Base Level Assessment
FYDP	Future Years Defense Program
IMA	individual mobilization augmentee
ISC-B	Integrated Security Construct–Scenario B
LCOM	Logistics Composite Model
MAJCOM	major command
MPA	Military Personnel Appropriation
MPES	Manpower Programming and Execution System
MSI	manpower standard implementation [code]
NAF	numbered air force
OAIT	organize, administer, instruct, and train
OARIT	organize, administer, recruit, instruct and train
OMB	Office of Management and Budget
OPM	Office of Personnel Management
OSD	Office of the Secretary of Defense
OSD CAPE	OSD Cost Assessment and Program Evaluation
PMAI	primary mission authorized aircraft
PBBE	planning, programming, budgeting, and execution
RC	reserve component
RED HORSE	Rapid Engineers Deployable Heavy Operational Repair Squadron Engineer
RI	reporting identifier
RMG	Readiness Management Group
RPA	remotely piloted aircraft
SDI	special duty identifier
SEI	special experience identifier

SMART Force	Strategic Manpower Assessment of Requirements–Total Force
TFA	Total Force Assessment
TFE	Total Force Enterprise
TPFDD	time-phased force deployment data
TR	traditional reservist
UIC	unit identification code
UMD	unit manning document
USC	United States Code
UTC	unit type code
WMP-3	War Mobilization Planning, Volume 3

1. Introduction

Quantifying total force manpower requirements in the U.S. Air Force involves multiple processes. The distribution of the Air Force's force structure across the active component (AC) and the reserve components (RCs)—the Air National Guard (ANG) and Air Force Reserve (AFR)—takes place in an integrated process managed by the Air Force Corporate Structure.[1] However, development of manpower requirements to support the force structure is less integrated, with active, ANG, and AFR requirements developed somewhat independently. Civilian and contractor workforces also support the missions of each component and are sized through their own processes. Additionally, there are discernible gaps between objectively derived manpower requirements and available manpower resources. As the Air Force encounters significant fiscal challenges in coming years, finding the best and most efficient fit of manpower to missions will become increasingly important. Meeting that challenge may require manpower processes that foster more effective integration of AC and RC organizations and more efficient use of workforces across the components.

The development and specification of qualitative personnel requirements is more integrated. Officer and enlisted classification guides, which describe the duties of and specify the required knowledge, skills, abilities, training, and education for various Air Force specialties, are common across AC and RC military workforces. A civilian classification structure, promulgated by the U.S. Office of Personnel Management (OPM), is used throughout the federal government.

In the research underlying this document, we explored policies regarding manpower and personnel requirements processes and the ways they appear to be implemented. We sought to understand the robustness of the various processes individually and the degree to which they complement each other and avoid over- or understating the requirements for various missions. Since some requirements are based on wartime needs rather than current or expected garrison needs, we sought to determine if a common-sight picture of wartime demands was used in various processes.

[1] The Air Force Corporate Structure consists of the Air Force Council (AFC), the Air Force Board (AFB), the Air Force Group (AFG), and the 13 mission and mission support panels (Air Force Instruction [AFI] 16-501, pp. 6–9). The AFC is chaired by the Air Force Vice Chief of Staff, and its membership consists of Air Staff and Secretariat principals (three-star and civilian equivalent, with select two-star members). The AFB provides flag-level (one- to two-star and civilian equivalent) review, evaluation, and recommendations within the same broad categories as the AFC. The AFG provides senior-level leadership (O-6 and civilian equivalents) a forum for initial corporate review and evaluation of appropriate issues within the same broad categories as both the AFB and AFC. The nine mission panels and five mission support panels review and develop options to present to the AFG. Through processes not documented in AFI 16-501, significant programming inputs to the corporate structure are made by major commands designated as *core function lead integrators*.

Force Structure Requirements Versus Manpower Requirements

Our research in this project did not address the distribution of force structure across the AC and RC.[2] The Air Force programs its AC/RC distributions within various weapon systems with help from a Total Force Enterprise model maintained and operated by the Air Force Directorate of Analyses, Assessments, and Lessons Learned (AF/A9) and a series of force composition analyses conducted by the Air Force Total Force Management Division (AF/A8XF) (documented in a forthcoming revision of AFI 90-1001). Actual AC/RC distributions are determined in part through leadership vectors and programming and budgeting decisions made within the Air Force Corporate Structure process and in part by congressional authorizations and appropriations that may reshape Air Force budget submissions (see, for example, Goulka and Englehart, 2013). Our research focused, instead, on the actions that accompany these processes—determining the manpower requirements to support the force structure as it is currently distributed, or programmed to be distributed, within the Future Years Defense Program (FYDP).

Scope of the Research

As discussed in Chapter Two, manpower processes differ across at least three dimensions:

- component (active, ANG, AFR)
- functional area (operations, maintenance, combat support)
- workforce (active military, active Guard/Reserve [AGR], air reserve technician [ART], dual-status technician [ANG equivalent of the AFR's ART], traditional part-time reservist/guardsman [TR], individual mobilization augmentee [IMA], civilian, contractor).

The permutations of these dimensions yield a large number of distinct manpower processes. Exploring all of them in detail would have been beyond the resources available for our research project. Instead, we sought early in the research to identify and focus on areas where requirement processes appeared to be problematic and that we found to be of greatest interest to our research sponsor. We deemed processes problematic if current practice appeared to depart significantly from policy or if processes appeared to be inadequately defined in policy, resulting in various stakeholders adopting potentially inconsistent or ineffective approaches.

Organization of the Document

Chapter Two provides an overview of the quantitative personnel requirements determination processes. Chapter Three provides a similar overview of qualitative processes. Chapter Four describes in more detail the practices we found in several illustrative functional areas. Chapter

[2] *Force structure*, in this context, denotes combat units and their associated equipment.

Five provides recommendations intended to promote greater coordination across components and functional areas and reduce slack requirements.

2. Quantitative Requirements Processes

In this section, we summarize the processes specified by the Air Force to quantify the manpower required to execute and support various missions. We also describe aspects of the current policies and processes that appear to be problematic.

Different Appropriations for Different Requirements

The manpower provided for an Air Force mission may include any combination of AC military, RC military, civilian, and contractor workforces. The processes for setting requirements for each type of workforce are somewhat disjointed, in part because they are funded through separate congressional appropriations. AC and RC military strengths are funded through centrally managed Military Personnel Appropriation (MPA) and Reserve Personnel Appropriation (RPA) accounts, respectively, and are subject to aggregate strength ceilings found in annual national defense authorization acts.[1] Civilian and contractor workforces are funded through Operations & Maintenance and other appropriations accounts (which are also used to procure goods and services) with considerable local discretion and much less central management control over how funds are expended.

Standard Processes for Quantifying Requirements

We found 11 distinct approaches to identifying manpower requirements for various segments of the total force. These approaches, discussed below, cover most contexts. Where none of these apply, Air Force commanders, functional managers, and manpower managers may rely on ad hoc approaches to identify requirements.

Manpower Standards

The most common approach to determining manpower requirements is the development and application of manpower standards. Standards are developed for specific work centers using management-engineering approaches outlined in Air Force Manual (AFMAN) 38-208. A standard generally includes an equation that relates required man-hours to one or more workload factors. The solution to a manpower equation is divided by an appropriate man-hour availability factor to calculate the aggregate manpower required for the work center. A standard also includes a *standard manpower table*, which provides a recommended disaggregation of the

[1] The ceilings are expressed in the form of "end strength"—the targets for officer and enlisted strengths on the last day of a fiscal year. Typically, some deviation around the specified end strengths is permitted by the legislation.

manpower requirement into grades and Air Force specialties (AFMAN 38-208, Vol. 1, pp. 90–92).

Most AC base-level support and logistics functions, as well as many unique activities, are covered by manpower standards. The standards may specify either military or civilian manpower or allow for substitution of civilian for specified military manpower. The standards also cover most full-time RC (technician[2] and AGR) manpower requirements, based on the workloads associated with maintaining an RC unit's equipment, supporting its garrison functions, and training its assigned part-time drilling reservists or guardsmen, referred to as *traditional reservists* (TRs) (Air Force Reserve Command Instruction [AFRCI] 38-201, p. 5; ANG Instruction [ANGI] 38-201, pp. 3–4).

Crew Ratios

For most AC and RC flying units, aircrew requirements are derived using approved *crew ratios* (the ratio of aircrews to primary mission authorized aircraft [PMAI]) and *crew complements* (the standard composition of crew positions on an operational aircraft) (AFI 38-201, p. 67). Exceptions are established for flying training instructor and student authorizations and other nonstandard requirements. Flying unit staff positions, such as those in wing headquarters or operational support squadrons, are governed by manpower standards.

Logistics Composite Model

Most AC aircraft maintenance manpower requirements are determined using the Logistics Composite Model (LCOM) (AFI 38-201, p. 82). LCOM is a simulation model that uses aircraft availability, maintenance policies, environmental factors, and probabilities of maintenance activities in conjunction with a defined operational scenario to determine maintenance manpower and other logistics requirements (AFMAN 38-201, Vol. 3, pp. 1–3).

Technical Estimates

Used primarily for medical requirements, technical estimates provide an alternative to traditional manpower standards for providing a workload-based estimate of manpower requirements.

Deployment Requirements

While most manpower requirements are predicated using garrison requirements and a standard (as opposed to wartime) military *man-hour availability factor*, requirements for TRs are, by policy, based on deployment requirements that cannot be met by available active duty manpower resources

---

[2] Throughout this document, we use the term *technician* to refer to AFRC's air reserve technicians (ARTs) and ANG's dual-status technicians.

(AFI 38-201, p. 46). Deployment manpower requirements are specified in various functionally specific *unit type codes* (UTCs) (see discussion below).

Management Headquarters Staffs

Management headquarters staffs (e.g., the Air Staff, major command (MAJCOM) staffs, or Air Force elements of joint and international organizations) have unique and changing requirements that have not been captured in standards or models. However, their sizes are subject to congressional and Department of Defense (DoD) scrutiny and are therefore closely managed (AFI 38-202; Department of Defense Instruction [DoDI] 5100.73).

Individual Mobilization Requirements

Similar to deployment requirements for RC units, IMA requirements are established predominantly for in-place wartime augmentation of AC units by individual part-time reservists. IMA requirements may also be justified on the basis of helping to meet recurring deployment requirements (AFI 38-204, p. 33). IMA positions are established at the request of and with justification provided by AC or non–Air Force organizations, subject to review and the availability of resources within the Air Force Reserve Command (AFRC) (AFI 38-201, pp. 55–56).

Other Processes

Service-type contracts provide an alternative source of manpower to meet Air Force mission and support requirements. They may be established through either conventional contracting processes or public-private competitions conducted under Office of Management and Budget (OMB) Circular A-76. When these contracts provide services that directly substitute for in-house Air Force functions, the Air Force tracks the contractor manpower equivalents that would be required to provide the service in-house (AFI 38-201, p. 20). Other processes that provide objective bases for manpower requirements include positions related to foreign military sales, treaties, and memoranda of understanding with activities outside the Air Force.

Requirements Process Coverage

Table 2.1 indicates the extent to which manpower authorizations of various types are covered by objective manpower requirements processes. Within the Manpower Programming and Execution System (MPES), a *manpower standard implementation* (MSI) code is associated with each authorization to indicate the process by which the manpower requirement was determined.[3]

---

[3] MPES is a database containing detailed information on each actual (funded) or potential (unfunded) manpower position within the Air Force. Primary responsibility for updating the detailed information rests with the MAJCOMs to which manpower has been allocated.

Appendix A contains a list of MSI codes, their mapping to the processes listed in Table 2.1, and a count of authorizations carrying each MSI code.

Table 2.1. Requirements Process Coverage by Workforce Categories (September 2012)

Workforce Category	Manpower Standard	Crew Ratio	LCOM	Technical Estimate	UTC	HQ Staffs	IMA	Other	No Standard Process	Total
Active Military	79,008	12,335	47,611	30,783	0	33,478	1	6,972	91,119	301,307
Active Civilian	25,257	103	754	15,490	0	19,459		9,219	79,652	149,934
AFR Civilian	2,043				65	508		4	1,175	3,795
ANG Civilian	754				0	1			833	1,588
AFR AGR	1,128	5		1	412	581	4		807	2,938
ANG AGR	3,725	90	39		7,590	74		1	3,488	15,007
AFR ART	3,825	60			40	210			6,449	10,584
ANG Tech	22,794	11			133	1		4	222	23,165
AFR TR	107	3,422	226		35,085	473			7,504	46,817
ANG TR	535	58	377		79,709	0		1	10,767	91,447
AFR IMA	1					19	8,669	1	21	8,711
Total	**139,177**	**16,084**	**49,007**	**46,274**	**123,034**	**54,804**	**8,674**	**16,202**	**202,037**	**655,293**

SOURCE: MPES.
NOTES: Includes funded authorizations only and excludes permanent party student authorizations. Some MSI coding errors are apparent in the data, such as citing manpower standards as a basis for TR positions or the IMA MSI for active military or AGR positions.

Figure 2.1 indicates the extent to which authorizations in each workforce category are covered by standard requirements processes. Within the collection of standard processes, we judged four processes (manpower standards, crew ratios, LCOM, and technical estimates) to be more rigorous than the others. These processes feature formulas or models whose inputs include workload drivers and workload-to-manpower relationships that have been derived empirically or through expert judgment, yielding objective, replicable results. The UTC process for determining TR requirements would be considered rigorous if it were consistently applied. As discussed below, however, application is inconsistent. Accordingly, we judged RC authorizations based on UTCs (carrying the UTC MSI code) as *potentially* rigorous (identified by crosshatched bar segments in Figure 2.1).

Figure 2.1. Rigorous and Standard Process Coverage by Workforce Categories (September 2012)

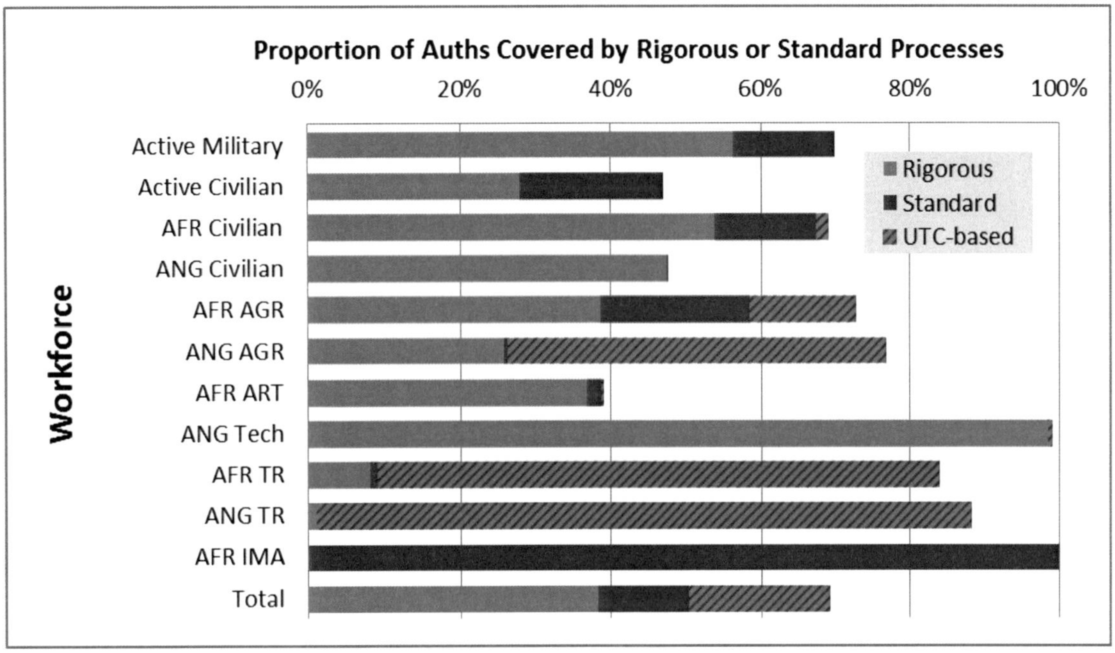

SOURCE: MPES.
NOTES: Includes funded authorizations only and excludes permanent party student authorizations. ANG Tech = ANG dual-status technician.

Age of Manpower Standards

Manpower standards provide the basis for more manpower authorizations than any other process. Their rigor, however, comes at a cost. Studies must be conducted at a relatively granular level and must be updated continuously to account for changes in organization, technology, and processes. Figure 2.2 indicates the distribution of the age of AC manpower standards, suggesting that the resources devoted to updating the standards using current practices are not equal to the task. Additionally, a review of manpower processes may be warranted to determine if they can be streamlined, perhaps by prudently trading off some level of measurement precision.

Figure 2.2. Age Distribution of AC Manpower Requirements Sources (April 2013)

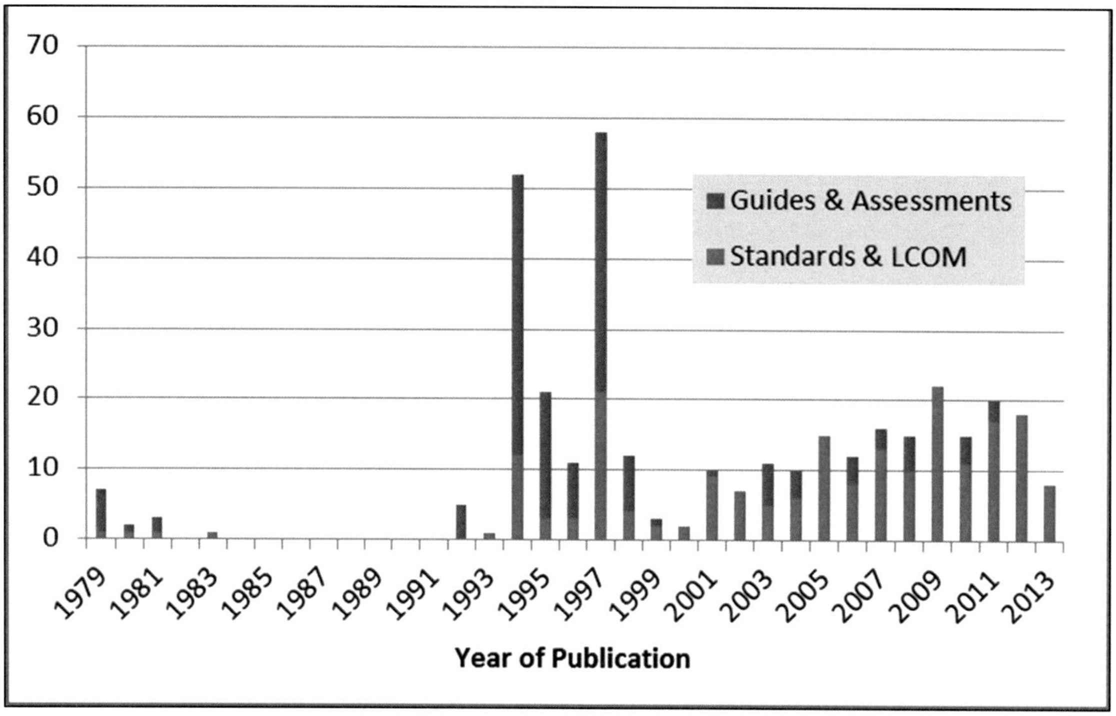

SOURCE: AF/A1MR.

Military Versus Civilian or Contractor Requirements

Air Force manpower policy requires the use of civilian or contractor manpower whenever military manpower is not essential (AFI 38-204, p. 20), presumably because civilian and contractor manpower is less costly than military manpower. For non–military-essential requirements, civilian manpower is required for inherently governmental activities and other specified conditions; for other positions, the relative cost of civilian and contractor manpower is the determining factor (AFI 38-204, Table 6.2).

Allocating and Extending Manpower Authorizations

While the standard approaches described above provide a means of identifying manpower requirements, additional decision processes govern which requirements are accepted as valid targets for resourcing. The Air Force Corporate Structure decision process used for force structure programming and budgeting (see description in Chapter One) also determines the aggregate manpower to be allocated to each MAJCOM and program element in each fiscal year through the FYDP. MAJCOMs extend allocated manpower to the level of detail needed for personnel management programs (grade, Air Force specialty, location, unit, current and future fiscal quarters, etc.) and record the extended authorizations in MPES.

In the Air Force Corporate Structure, AFRC is sometimes treated as a MAJCOM during programming and budgeting deliberations for force structure. However, both the ANG and AFRC are responsible for programming and budgeting for their own manpower and allocating that manpower to their units. In this process, the Air Force Directorate of Manpower, Organization, and Resources (AF/A1M) is responsible for arbitrating disagreements (AFI 38-204, pp. 1–12).

Allocated and/or extended authorizations may be greater than or less than requirements identified through standard approaches. If identified requirements exceed allocated aggregate manpower, MAJCOMs may record the unfunded requirements through appropriate entries in MPES.[4] Table 2.2 indicates the funded proportion of authorizations in each workforce category and manpower process as of September 2012. In total, AC military and civilian workforce categories have the lowest percentage of funded authorizations, while AFR/ANG TRs have the highest.

Table 2.2. Percentage of Total Authorizations That Are Funded (September 2012)

Workforce Category	Manpower Standard	Crew Ratio	LCOM	Technical Estimate	UTC	HQ Staffs	IMA	Other	No Standard Process	Total
Active Military	87.7	99.0	95.7	98.6		95.1	100.0	92.7	89.6	91.9
Active Civilian	86.8	100.0	96.1	93.7		94.3		84.0	91.7	90.9
AFR Civilian	98.9				92.9	99.0		100.0	98.7	98.8
ANG Civilian	86.4					100.0			99.8	92.9
AFR AGR	94.7	100.0		100.0	99.8	94.6	100.0		96.5	95.9
ANG AGR	99.0	100.0	100.0		100.0	98.7		100.0	99.6	99.6
AFR ART	91.3	100.0			66.7	75.8			96.6	93.9
ANG Tech	81.7	100.0			57.8	14.3		44.4	44.4	80.9
AFR TR	111.5	100.0	100.4		101.6	146.0			104.1	102.2
ANG TR	100.0	100.0	100.0		100.0			100.0	100.0	100.0
AFR IMA	100.0					90.5	75.0	100.0	100.0	75.1
Total	87.1	99.2	95.8	96.9	100.4	95.0	75.0	87.6	91.8	92.9

SOURCE: Data extracted from MPES.

[4] Friction in the system generally leaves disparities between the counts of authorizations allocated by the Air Staff and authorizations extended by the MAJCOMs in MPES. The disparities tend to be greater in the latter years of the FYDP, which naturally receive less attention from MAJCOM manpower staffs.

Garrison Versus Deployment Requirements

The processes described above largely pertain to garrison requirements. Individual manpower authorizations are aggregated into unit-level building blocks, generally on the scale of a squadron or distinct activity. Each unit-level building block has a distinct unit identification code (UIC). Locally, the manpower authorizations in a UIC are listed on a unit manning document (UMD). At the Air Force and MAJCOM levels, authorizations are listed in MPES.

For wartime planning and execution, the aggregate building block is a UTC, which is typically on the scale of a defined functional capability that can be drawn from a single garrison unit.[5] Air Force and MAJCOM functional area managers (FAMs) establish UTCs to represent the capabilities needed in wartime contingency planning scenarios or to meet crisis action planning demands (generally in the context of ongoing rotational deployment requirements). As Figure 2.3 illustrates, a UTC is postured by mapping its positions to positions within a UIC, i.e., identifying which garrison positions' incumbents would deploy if the UTC were tasked for deployment. In the posturing process, MAJCOM FAMs and commanders are required to insure that the maximum numbers of funded authorizations assigned to the UMD are postured in standard UTCs, that any residual deployable authorizations not in standard UTCs are postured into associate UTCs (A-UTCs), and that Air Force manning positions are not assigned to more than one UTC record (AFI 10-403, p. 87). Thus, for AC units, garrison UICs determine the resources available to be postured in UTCs. However, for TR requirements, the process is intended to work in the opposite direction—UTC requirements, by policy, determine the required number of TR authorizations to be established in UICs. RC garrison requirements, including the training of TRs, are, by policy, fully met using full-time workforces, consisting of technicians, AGRs, and civilians. Generally, TR UIC requirements are sized to meet total Air Force UTC requirements that exceed the UTCs postured using available AC manpower.[6] For the support functions we examined (see Chapter Four), in practice we found little connectivity between the AC's determination of UTC requirements, the posturing of UTCs in the RC, and hence the sizing of TR requirements in RC UICs. This limited connectivity is represented by the dashed line joining AC and RC processes in Figure 2.3.

---

[5] A UTC is a Joint Chiefs of Staff–developed and –assigned code consisting of five characters that uniquely identify a type of unit (Chairman of the Joint Chiefs of Staff Manual [CJCSM] 3122.01 series). As used in the Air Force, a UTC is generally a modular part of a squadron-sized unit identified by a UIC.

[6] As an exception to this general principle, aircrew TR UIC requirements are sized using crew ratios rather than UTC requirements.

Figure 2.3. Garrison and Deployment Requirements Processes

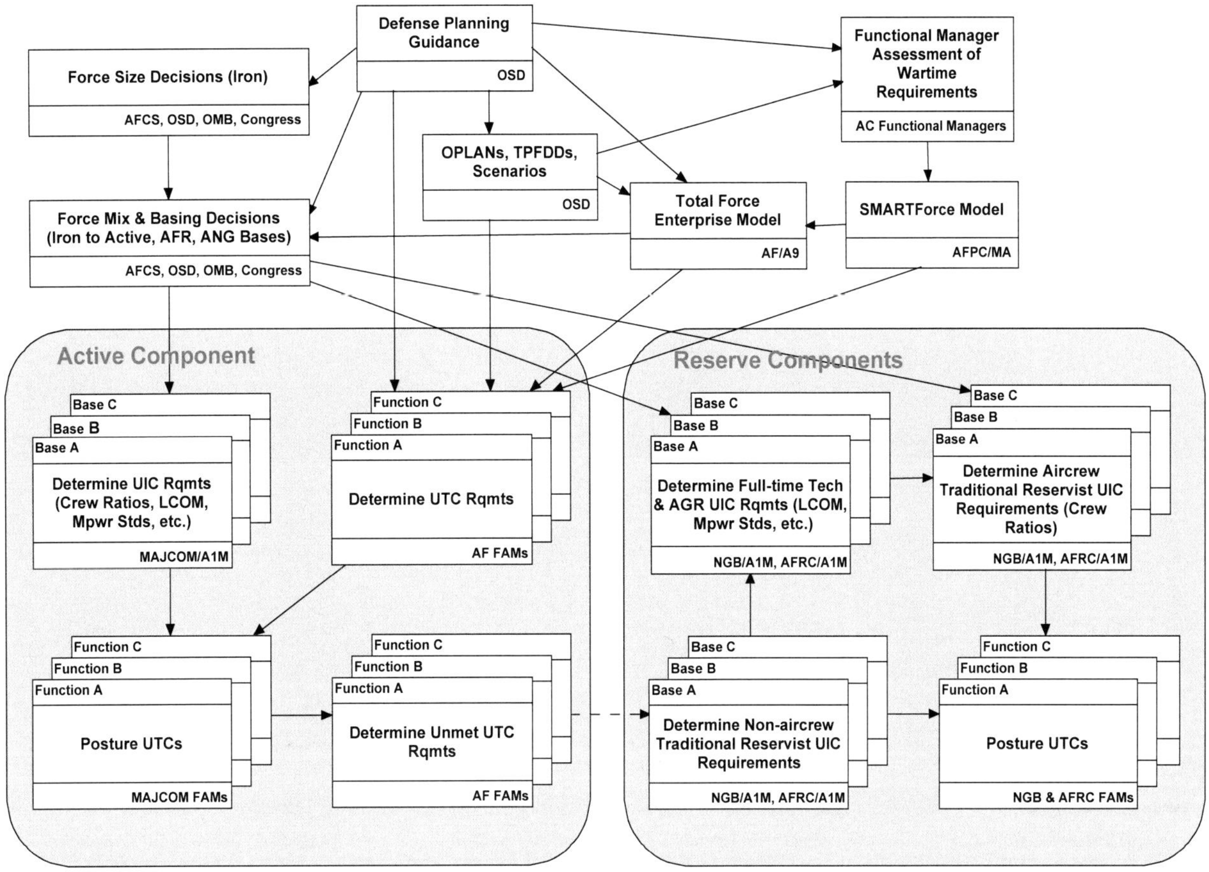

NOTES: AFCS = Air Force Corporate Structure; OSD = Office of the Secretary of Defense; OPLAN = operation plan; TPFDD = time-phased force deployment data; AFPC/MA = Air Force Personnel Center Directorate of Manpower; NGB/A1M = National Guard Bureau, Manpower, Organization, and Resources Division; AFRC/A1M = AFRC Manpower, Organization, and Resources Division.

Issues with Quantitative Requirements

In this section, we evaluate the processes described above to distill issues that appear to warrant further attention.

Manpower Requirements Workforces

Insufficient capacity to develop and apply standard processes is evident in some of the material presented above. As indicated in Table 2.1 and Figure 2.1, significant portions of some workforce types are not covered by standard processes. Further, Figure 2.2 indicates that many manpower standards are old enough to cast doubt on their applicability to current processes, many of which may have gained efficiencies due to technological advances.

Some officials have suggested that capacity limitations are not related to the number of personnel engaged in requirements processes (i.e., the workforces in the Air Force's five manpower requirements squadrons [MRSqs]), but rather to the composition of those workforces. In this view, the squadrons employ too few management engineers, operations researchers, or other individuals with the quantitative skills required to design, manage, and execute complete end-to-end manpower studies. Table 2.3 indicates that 140 of the 252 authorizations (55.6%) in these units are officers or comparable-grade civilians. Thus, the proportion of the workforce that can be expected to have four years of college education is relatively high. Nonetheless, managers responsible for oversight of these units might want to examine the credentials of this workforce more closely, particularly the 1st and 5th squadrons, which are relatively lean in their proportion of officers and comparable-grade civilians.

Table 2.3. Grade Distributions in Manpower Requirements Units

Grade	1st MRSq	2nd MRSq	3rd MRSq	4th MRSq	5th MRSq	6th MRFlt	7th MRFlt	Total
O-5	1	1	1	1	1			5
O-4	2	1	2	1	1	1	1	9
O-3	1	2	1	2	2			8
O-2	1	2	1	2				6
GS-14							1	1
GS-13	2	4	4	3	3	1	1	18
GS-12	7	17	13	7	5	1	4	54
GS-11	7	10	6	8	3	1	4	39
Subtotal	21	37	28	24	15	4	11	140
Percentage of Total	47.7	66.1	57.1	53.3	40.5	66.7	73.3	55.6
E-8	1	1	1	1	1			5
E-7	6	6	4	4	6	1	1	28
E-6	8	8	9	9	9	0	2	45
E-5	7	3	7	7	5	1	1	31
GS-06	1							1
GS-05		1			1			2
Subtotal	23	19	21	21	22	2	4	112
Total	44	56	49	45	37	6	15	252

SOURCE: MPES, September 2012.

Unconsolidated AC and RC Requirements

The Air Force's sponsorship of the research underlying this report was motivated in part by the objective of consolidating collocated AC and RC manpower, personnel, and services activities, thereby improving service and reducing total force manpower requirements. (In some cases, as many as three missions support squadrons or flights [one each for active, ANG, and

AFR units] that may be operating at the same location.) Our research revealed that fully realizing this worthwhile objective requires key changes to legislative restrictions on the utilization of RC manpower and to Air Force manpower requirements policies and processes.

Legislative Restrictions on the Utilization of RC Manpower

Reserve forces can operate in either training or operational modes. The operational mode is typically in support of named operations, where forces are made available to combatant commanders (COCOMs) to perform missions in support of national security needs.[7] The RC operates under a regulatory structure that limits the circumstances under which and numbers of personnel that can be utilized to serve in operational missions. Technicians performing their primary duties (organizing, administering, instructing, and training the RCs—the so-called OAIT duties) and AGRs performing similar primary duties (organizing, administering, recruiting, instructing, and training—the so-called OARIT duties) are limited by statute and appropriations language from duties "supporting operations or missions" if they would interfere with the performance of those primary duties. Similarly, RPA funding for TRs is specifically dedicated to "performing drills or equivalent duties." With limited exceptions, only activated reservists, paid using MPA funding, may be used to perform operational missions.[8]

A summary of how AC and RC workforces may be used to meet various requirements can be found in Table 2.4. The restrictions on duties of technicians and AGRs are explicit. Per 32 United States Code (USC) 709 (for the ANG) and 10 USC 10216 (for the AFR), the primary duties of technicians must be OAIT duties or the maintenance of RC equipment. Technicians may provide support to federal operations or missions only if it is incidental to or does not interfere with their primary duties. The use of AGRs is similarly restricted to OARIT duties under 10 USC 12310 and 10 USC 101(d)(6)(A).

---

[7] The strategic airlift portion of mobility air forces typically operates on a daily basis on missions that are not explicitly linked to named operations.

[8] Anecdotally, we hear that workarounds to these restrictions have been employed. However, also anecdotally, we hear from individual reservists and commanders that the restrictions inhibit the straightforward use of RPA-funded technicians, AGRs, and TRs in operational missions.

Table 2.4. Restrictions on the Utilization of Various Manpower Types

	Train	Support and Administer RC	Operate	Deploy	Notes
Active Military	X		X	X	
Civilian	X	X	X	X	
AGR	X	X	X	X	Must not interfere with primary duties (OARIT)[a] or must be in support of a mission assigned to the RC
ART/dual-status technician	X	X	Man-days	Man-days	Activation must not interfere with primary duties (OAIT) or the maintenance of equipment[b]
TR	X		Man-days	Man-days	When not activated, generally restricted to training and other specific RC-related duties[c]
IMA	X		Man-days	Man-days	Assigned against positions in the AC (AFI 36-2629)

NOTES: "Man-days" denotes that the indicated manpower type generally must be activated and provided with MPA man-days in order to be used for operations or deployments in support of active missions.
[a] References: 10 USC 12310, 10 USC 101, 32 USC 328.
[b] References: 10 USC 10216, 32 USC 709.
[c] Restrictions embedded in annual MPA and RPA appropriations, 13 USC 1301 (the "Purpose Statute"), and the Anti-Deficiency Act.

Restrictions on the use of drilling reservists are not statutory, but are rooted in annual RPA language, which usually stipulates that RPA funding must be used for duties specified in 10 USC 12310 (organizing, administering, instructing, and training RC personnel), for other duties specific to administration of the RC (10 USC 10211, 10305, 8030, and 12402 and 32 USC 708 or 502[f]), or "while undergoing reserve training, or while performing drills or equivalent duty or other duty" (see, for example, Title I, Military Personnel, in Public Law 112-10, the Department of Defense and Full-Year Continuing Appropriation Act, 2012).

The effect of these restrictions is that RPA-funded manpower cannot be used to perform regular and recurring duties that are part of an active mission, including garrison support of active units. Except for incidental contributions, activity devoted to non-training, operational-mission activities must be performed by activated reservists compensated using MPA man-day funding. RPA-funded manpower includes the 39 days of a TR's time available as monthly drill periods and annual training. It also includes the normal duties of technicians and AGRs, although technicians and AGRs can be activated and paid from MPA funds.

These restrictions are apparently intended to insure that RC units and individuals receive sufficient training to remain ready for mobilization. For flying units in the Air Force, and similarly for combat-arms units in the other services, continuation and upgrade training is an important part of the mission when not deployed or employed operationally. For combat air forces flying units in particular, continuation training is the primary mission of the unit while in garrison. The statutory restrictions discussed here seem to have been formulated with these kinds of units in mind.

But continuation training is not common in non-flying Air Force units, such as the mission support squadrons and flights that house local manpower, personnel, and services activities. While personnel in non-flying units receive various forms of ancillary training, skill upgrade training, and professional military education, it is much more limited in its time demands than the continuation training required in flying units. For active duty military personnel on standard 40-hour workweeks, Air Force manpower standards set aside four hours per month for training and education (AFI 38-201, Table A3.1), or the equivalent of six days per year. By comparison, the 39 days per year earmarked by appropriations funding for the training of TRs appears to be an order of magnitude greater than needed for those in non-flying units, particularly those in agile combat support (ACS) specialties.

Because of these restrictions, it is difficult to use RC personnel cost advantageously as part of a consolidated team supporting the garrison needs of local AC and RC units. This issue is particularly pertinent in garrisons where associated AC and RC units share a common mission and equipment. Commanders are legally compelled to insure that the 39 days of drill and annual training are devoted predominantly to training and make only incidental contributions to active missions. To maximize the benefits of the RC as an operational reserve, legislative changes are needed to ensure that time and effort funded from reserve appropriations (for technicians, AGRs, and TRs) can be used freely to support active missions in garrison.[9] Without legislative relief, the broader use of RC personnel as part of a consolidated activity serving active mission needs is either legally suspect (if RPA-funded manpower is widely used to support active missions) or unnecessarily expensive (if RPA-funded manpower is used only incidentally to support active missions).

Manpower Requirements Policies

To realize manpower savings through the consolidation of AC and RC support functions, changes in the policies and processes for both AC and RC manpower are needed. At the local level, AC manpower requirements in support functions are based on AC garrison workloads. On the RC side, technician and AGR requirements are based on RC garrison support and training workloads, but TR requirements should be predicated on providing UTCs to meet Air Force–wide deployment requirements. In associate units, efficiencies would likely be realized if AC, technician, and AGR manpower standards were consolidated and modified to use total force missions as workload drivers (including both AC and RC garrison support needs and the training of TRs) and to allow the blending of AC, technician, AGR, TR, and civilian manpower to meet local needs.

---

[9] We would expect alternative legislation to continue to specify that reserve units and individuals retain some level of readiness for mobilization. This could be accomplished by specifying the outcome—a level of readiness—rather than the input—resources devoted to training—presumed to be necessary to maintain sufficient readiness.

Weak Links in Determining Wartime Requirements for ACS

Our interviews with manpower managers (Air Force Manpower Requirements Division [AF/A1MR]), the Air Force War Planning and Policy Division [AF/A5XW]), and various functional managers revealed multiple weak links in the process of determining wartime UTC requirements for ACS functions.

As described above, UTC requirements, by policy, provide the foundation for TR requirements in most functional areas (all except aircrew). Functional managers should be assessing deployment demands, determining the degree to which they can be met by available AC forces, and establishing requirements for TRs to meet the balance of demands. Demands at a specific deployment location depend on the existing base infrastructure; the number, type, and mission of the aircraft bedded down; the total base population; and the level of conventional and unconventional threats to which the base is exposed (Snyder and Mills, 2004, pp. 7–11). These demands, in turn, are affected by the scenario selected and vary across the peacetime, surge, and post-surge phases of a contingency operation. During peacetime and post-surge phases, the availability of resources to meet demand is also a function of varying deploy-to-dwell ratio policies.

We found, first, that there is no rigorous, scenario- and TPFDD-based wartime planning process to which functional managers can turn to find accepted statements of demand for ACS.[10] A useful statement of demand would specify the number of deployment locations that must be supported, along with key characteristics of each location—the number and type of aircraft planned for the location, the number of personnel planned for the location, the threat environment of the location, and the assumed infrastructure at the location. Further, the statement of demand would include, for each location (either specific or generic) a list of the types and numbers of ACS UTCs required at the location.

Planning for operational force structure uses AF/A9's Total Force Enterprise (TFE) model to determine demands across the phases of contingency operations for OSD-approved contingency scenarios. As currently configured, however, the TFE model does not estimate demand for maintenance, support, and other ACS functions. Rather, ACS requirements derived from AFPC/MA's Strategic Manpower Assessment of Requirements–Total Force (SMARTForce) model are used as an input to the TFE model. SMARTForce, in turn, relies on functional managers' assessments of required wartime UTCs, which are based on a specified set of wartime deployment locations and their assumed characteristics (AFPC/MA, 2012).

Second, we found no system of record in which functional managers can represent their total force wartime requirements, which, to be useful for manpower planning purposes, would be expressed in terms of the count of each UTC required to support the accepted set of scenarios in

---

[10] Assessments of the supply and demand of manpower resources for deployment in non-operational functions were last conducted in 1995 (a Force Sizing Exercise [FORSIZE]/Base-Level Assessment [BLA]) and 2000–2001 (a Total Force Assessment [TFA]) (AFPC/MA, 2012).

surge and post-surge phases, taking into account the need for deployment rotations in post-surge phases and including a breakout of the number of UTC rotations to be filled by the AC and RC. Without some such system of records, AFRC/A1M and ANG Manpower, Organization, and Resources Division (ANG/A1M) have no enduring point of reference to use in sizing TR strengths in ACS functions. As of the time of this writing, AF/A1M and AFPC/MA sources indicated that the UTC requirements aspect of this system can be derived from SMARTForce and that they will take steps to post it so it can be available to functional managers. However, further development would be required to include an AC/RC breakout of required UTCs.

Consistent with our findings, other research conducted within Project AIR FORCE has found that wartime ACS capacities vary widely across functional areas (Mills et al., 2014). Figures 2.4 and 2.5, developed in that research, indicate the number of generic deployment bases that can be supported by twelve functional areas in surge and steady-state phases of wartime scenarios. The wide range of capacities strongly suggests that some functions are over-resourced while others may be under-resourced. Over-resourcing creates the unnecessary expense of slack resources. Under-resourcing creates deployment stress and limits the total number of deployment locations that can be supported with available ACS resources. Moreover, it is apparent from these data that either the responsible functional managers have radically different ideas concerning wartime requirements or the ANG and AFRC are not sizing TR requirements to complement AC warfighting capacity.

We examined deployment patterns of AC and RC personnel during a recent decade (fiscal years 2002–2011) to determine if wartime demands during what is regarded as a relatively demanding period seemed to justify the non-maintenance ACS TR resources maintained in the RC.[11] Table 2.5 shows these deployment demands in each component by broad mission area, as indicated by the first digit of the individual's duty Air Force specialty code (AFSC).[12] The average count of assigned military personnel in each mission area is indicated, along with the proportion of that count that, on average, was deployed at any one time.

---

[11] *ACS* is the term used by the Air Force to denote non-operational support functions. In some contexts it includes aircraft and missile maintenance functions and in others it does not. In this document, we exclude these maintenance workforces from our definition of ACS.

Our data are derived from a RAND-maintained dataset that defines *deployment* as time away from the home station in support of COCOM taskings. RAND's deployment dataset is compiled from military pay files, contingency management files, and other sources obtained from the Defense Manpower Data Center (DMDC).

[12] The definition of *ACS* we have adopted for use in this document includes mission areas 3 through 7 in Table 2.5.

Figure 2.4. Surge Capacity by Functional Area

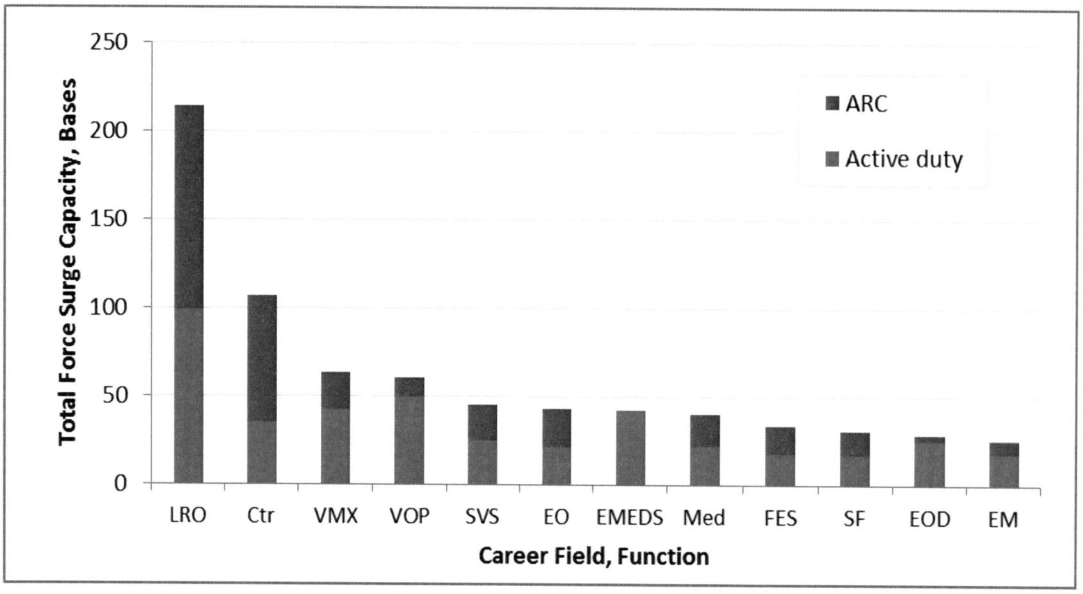

SOURCE: Mills et al., 2014, Figure 3.1.
NOTE: Career fields or functions are as follows: SF = security force; TMO = traffic management ops; EO = engineering ops; Ctr = contracting; LRO = logistics readiness officer; FES = fire emergency svc; VMX = vehicle maintenance; EM = emergency mgt; EOD = explosive ord disposal = VOP: vehicle ops; EMEDS = expeditionary med spt; SVS = services.

Figure 2.5. Steady-State Capacity by Functional Area

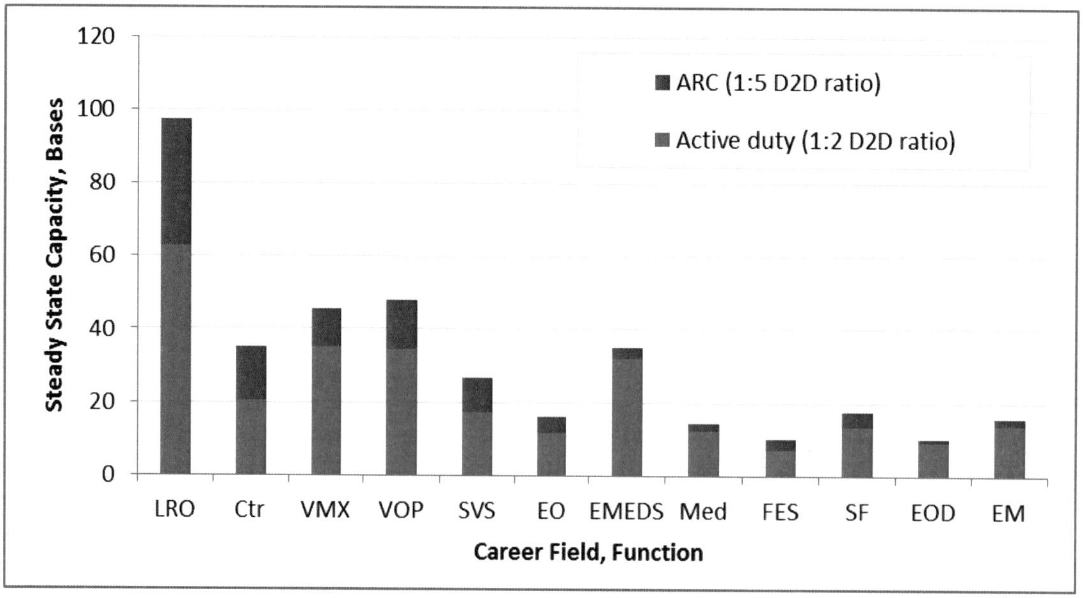

SOURCE: Mills et al., 2014, Figure 3.2.

Table 2.5. Rates of Deployment, by Component and Mission Area, FYs 2002–2011

	Active		AFRC		ANG	
Mission Area	Average Count	Average Percentage Deployed	Average Count	Average Percentage Deployed	Average Count	Average Percentage Deployed
1 - Operations	58,197	8.2	10,272	5.6	13,335	4.4
2 - Maintenance	105,445	6.5	23,536	2.2	38,885	2.4
3 - Mission Support	79,662	8.8	16,693	2.2	29,116	2.9
4 - Medical	29,992	3.8	9,641	1.3	5,749	1.4
5 - Chaplain/Legal	2,923	5.5	1,254	0.8	887	1.2
6 - Financial/Acquisition	10,651	4.4	1,276	0.6	1,371	0.7
7 – Special Investigations	1,148	10.5	327	1.6	0	–
8 - Special Duty	7,508	3.4	1,285	1.1	1,331	1.4
9 - Reporting ID	16,038	0.9	446	1.5	5,836	0.2
Total	311,564	6.7	64,729	2.5	96,510	2.6

SOURCES: DMDC and Air Force (multiple files).

Ostensibly, these deployment rates could be evaluated by comparing them to deploy-to-dwell ratios set by policy: a 1:2 deploy-to-dwell ratio for the AC (33 percent deployed at any one time) and a 1:5 mobilization-to-dwell ratio for the RC (16.5 percent deployed at any one time).[13] But those ratios are intended to apply to available populations in units designated for deployment rather than to total populations. To estimate an appropriate benchmark, we modeled a notional ACS functional area with 1,000 AC garrison requirements and then estimated the cost-minimizing mixes of active military, civilian, and part-time reserve authorizations at various levels of planned wartime requirements.[14] As illustrated in Figure 2.6, the model indicates that RC assets enter the cost-minimizing mix only after planned wartime requirements are above 140,

[13] Deploy-to-dwell policies were issued by the Secretary of Defense in a 2007 memorandum. The memorandum specified a 1:5 mobilized-to-demobilized ratio for Guard and Reserve *units* and a similar 1:2 deployed-to-home station ratio for active *units*. The 1:5 mobilization-to-demobilization policy is captured in DoD Directive 1235.10 (p. 6), where it is expressed as a limit for both *unit* and *individual* mobilization frequencies. In recent periods, Air Force practice has tended toward deployments of UTCs that are much smaller than conventional units. Consistent with that practice, AFI 10-402 seems to apply the 1:5 and 1:3 ratios as limits on involuntary individual rather than unit mobilization and deployment. A safe assumption might be that the limits apply to both individuals and units, but that assumption leads to conclusions that would overly constrain force availability. A unit that has sufficient dwell time for a prospective mobilization or deployment may have individual members who do not have sufficient dwell time. Likewise, individuals with sufficient dwell time may be assigned to units that do not have sufficient dwell time. We were unable to find OSD or Air Force guidance that resolves the unit-versus-individual ambiguities in the policies and we noted in conversations with various Air Force functional managers and deployment planners that understandings of the policy vary markedly among those responsible for tracking and enforcing it. Accordingly, in some places in this document we apply or discuss the policy as a unit limit and in other places as an individual limit.

[14] This model provides a useful estimate of a cost-minimizing force mix in functions for which personnel costs are the primary driver of overall costs. It is not applicable in, for example, evaluating operational force mixes, where many other cost categories are significant.

or 14 percent of the AC garrison requirement. Below that level of wartime demand, the cost-minimizing mix includes AC military and civilian personnel, but no part-time RC personnel. Above that level, it includes AC and part-time RC military personnel, but no civilians.

Figure 2.6. Least-Cost Manpower Mix in Notional ACS Function

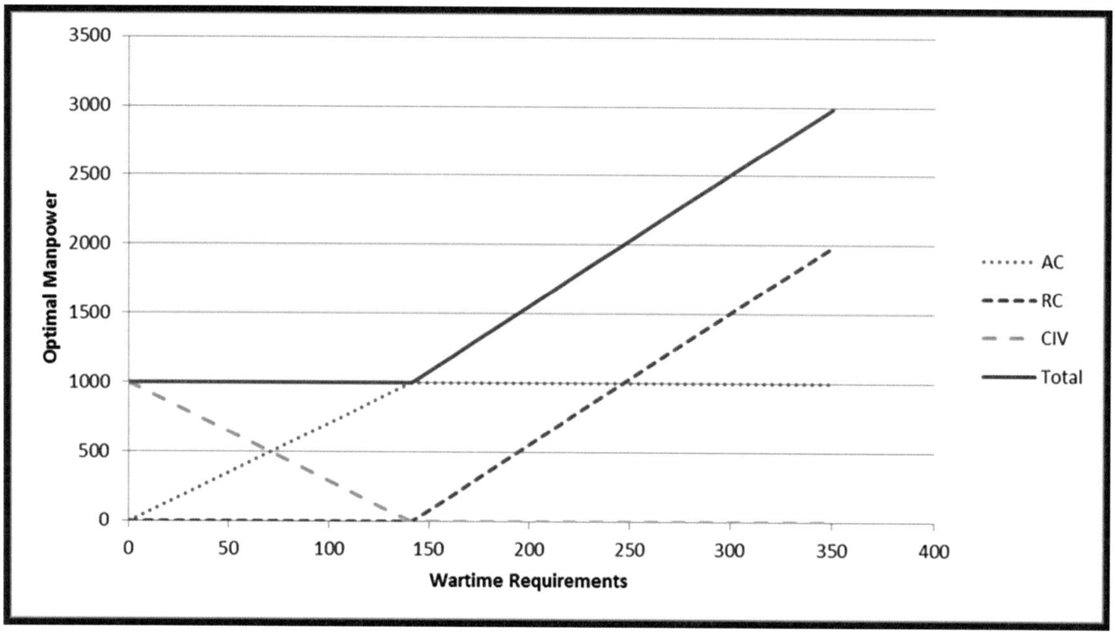

SOURCES: Deployment availability rates used in the model (42% for the AC, 63% for the RC) derived from AF/A1MR categorizations of current manpower and the historical average percentage of individuals not available for deployment due to medical, family, disciplinary, and other reasons. Deploy-to-dwell rates (1:2 for the AC, 1:5 for the RC) reflect DoD guidance. Cost data (annual cost of $90,918 for the AC, $39,717 for the RC, and $80,500 for civilians) are derived from AFI 65-503 FY 2012 standard composite rates.
NOTES: Deployment availability rates consider individual availability, as well as the proportion of authorizations in institutional, in-place warfighting, and other similar non-deployable activities. RC cost includes TRs, as well as the AGRs and technicians required to train them. Civilian cost is adjusted so that the civilian officer-equivalent/enlisted-equivalent grade composition equals the military officer/enlisted grade composition.

The data in Table 2.5 indicate that all ACS and maintenance mission areas (first digits 2 through 6) are, in the aggregate, comfortably below the 14 percent threshold estimated using our cost-minimizing model. While individual specialties may (and certainly some do) exceed this threshold, indicating that they may need support from the RC to meet wartime demand, the majority of ACS specialties do not appear to need RC support.

Consistent with this analysis, the data in Table 2.5 indicate that the RC has been called upon sparingly to provide ACS wartime support, even during a period of relatively heavy deployments. Excluding the Operations mission area, the highest average proportion of the RC workforce deployed at any one time was 2.9 percent.[15] RC ACS TR personnel costs are currently

[15] As discussed above in a section on garrison vs. deployment requirements, the TR workforces should be sized to support UTC requirements that cannot be met using postured AC UTCs and assuming an acceptable deploy-to-dwell

$1.4 billion per year.[16] Assuming that improved UTC-based analyses, which would also include more canonical major combat operations, confirm the very limited need for these resources, considerable savings are possible—even greater than $1.4 billion if full-time positions required for the administration and training of the TRs are also eliminated or if excess TRs in aircraft maintenance functions are also considered. These savings could be realized with no change in the RC aircraft inventory, since garrison support and the maintenance of the aircraft inventory is provided by full-time RC workforces. In an era of budget-driven force reductions, TR ACS authorizations are the proverbial "low-hanging fruit."

Tenuous Linkage of IMA Authorizations to Wartime Requirements

Historically, IMAs were assigned to key command positions for training and availability in case of mobilization, but there has been a shift in the usage of IMAs in today's environment to include augmentation in other than key command positions and volunteering for deployment in support of active duty missions. IMAs are defined as "individual members of the Selected Reserve assigned to a Reserve component (RC) billet in an active component or non-DoD organization. IMAs are trained individuals pre-assigned to billets that must be filled to support mobilization requirements, contingency operations, operations other than war, or other specialized or technical requirements" (DoDI 1235.11, Enclosure 2). The current criteria for validating IMA positions, contained in AFI 38-201 and AFI 38-204, are as follows:

- **Mobilization.** A federal active duty status in which airmen are involuntarily called to active duty. Mobilization IMA authorizations should match the grade and AFSC of the active duty billet to be backfilled or augmented.
- **Contingency operations.** IMAs support requirements related to Air and Space Expeditionary Forces (AEF); backfill active duty positions that deploy from an AEF location; fill a recurring AEF deployment requirement; or augment Air Staff, MAJCOM, and numbered air force (NAF) staff functions that must expand operations. Contingency operation IMA requirements must reduce the operations tempo of the active force, provide workload relief for units that lose personnel on steady-state deployment requirements, or augment MAJCOM or NAF staffs not manned to handle the expanded hours and they should match the active duty billet to be backfilled or augmented.
- **Specialized, technical, or scientific.** IMAs augment active forces when the workload is mission essential and to maintain a military capability requiring specialized, technical, or scientific knowledge or experience not otherwise available in the active duty force.
- **Economic.** IMAs augment active forces when the workload is mission essential and the cost of IMA augmentation is less than the alternative manpower resources. To meet the economic criterion, funding for the new IMA position must be offset by funds from

---

ratio. But since neither wartime UTC requirements nor deploy-to-dwell ratio calculations have not been well defined, we (and the Air Force) are unable to precisely estimate how large the TR workforces need to be.

[16] This figure is derived from a weighted average of standard composite rates for officer and enlisted TRs in the ANG and AFRC multiplied by a count of 75,000 TR resources in ACS functions.

existing programs or funds that will be requested in the planning, programming, budgeting, and execution (PPBE) process.

- **Management and training of AFR personnel.** Partial justification of an IMA authorization can be the management and training of AFR personnel, provided that the IMA's proficiency in his or her primary responsibilities will not be degraded, IMA use is economical, and the IMA will exercise the leadership and management skills required of the position and rank.

Although these criteria are broad, the directives indicate that IMAs are not intended to augment AFRC force structure or ANG units, replace temporarily unavailable military personnel or civilian employees who are ready reservists of the military services, or manage and train AFR personnel when that is the sole justification for an IMA requirement (AFI 38-201, AFI 38-204).

An issue unique to IMAs is that they are incorporated into active units, with resulting ambiguity regarding whether the AC or the RC is responsible for their development and utilization. To aid in the management of IMAs, the Readiness Management Group (RMG) was created within AFRC in 2005 to maintain oversight and administrative control of the IMA force. The RMG has a role in educating active duty supervisors about their responsibilities for managing, training, and developing IMAs. Closer management of IMAs is provided by RMG's program managers or by designated central and single program managers for selected functional communities (see Table 2.6).

Table 2.6. IMA Authorizations by Program Manager (September 2012)

Program Management	Officer	Enlisted	Total
Central Manager			
Chaplain (AFRC/HC)	270	97	367
Legal (AFRC/JA)	567	181	748
Medical (AFRC/SG)	494	70	564
Single Manager			
Intelligence (AFISRA) (see note below)	342	218	560
Special Investigations (AFOSI)	162	243	405
RMG Program Managers	3,236	2,831	6,067
Total	5,071	3,640	8,711

SOURCE: MPES.
NOTE: Although applicable directives (AFI 38-201, AFI 38-204) and MPES records indicate that AFISRA IMAs are centrally managed, AFISRA IMAs have been decentralized and are now managed through the RMG program manager structure.

Requests for IMA authorizations are initiated by potential using organizations in the AC, coordinated with the using organization's major command or the centrally managed/singly managed program manager, and forwarded to AFRC/A1 for review, coordination, validation, and approval. Each IMA position is associated with an active duty billet, with the IMA position assuming the same grade and AFSC, but AFRC programs and budgets the total number of funded IMA billets. An annual IMA program review by the using command or central manager/single manager determines if IMA requirements are still valid, considers new and unfunded requirements, and creates a prioritized list of IMA billets. If a requirement has been unfunded for three years, it is scrutinized further and potentially taken off the UMD. Additionally, AF/A9 provides an IMA Authorization Tool (IMAAT) to aid in prioritizing requirements. The tool's inputs include requester priorities, critical AFSCs, stressed career fields, and projected career-field shortfalls and surpluses.

Sneed and Kilmer (2012), used Beer's (1972) viable systems model to analyze the IMA program and recommended the following: determine and communicate the role of IMAs in the Air Force, add a strategic planning mission to the responsibilities of the Commander, AFRC, and improve the communication and information channels used by the IMA program. They argue that, with these changes and the flexibility allowed by policy, the IMA program could become an operational reserve rather than remain in its current ambiguous status.

AFRC appears to be properly engaged in the management of IMA personnel, but has more voice than is appropriate in the management of IMA positions. Using commands have a better sense than AFRC does of where augmentation and backfill resources would be most useful. Consistent with Sneed and Kilmer's (2012) recommendations, we recommend migration to a system in which aggregate allocations of IMA manpower are made by the Air Staff to using commands, similar to the allocation of other manpower resources, as part of the Air Force corporate manpower programming process.

Utility as Mobilization Augmentees

We have heard from users of IMA positions that their role in augmenting headquarters staffs during periods of intense activity is an important and useful one. It is less clear that the program is useful for backfilling the specific active duty positions with which IMA positions are associated. For this approach to be effective, the IMA position must first be associated with an active position that is tapped for deployment or heavily tasked during contingencies. Given the large number of active positions subject to deployment and the relatively low number of IMA positions, the probability that an active position will have a designated IMA backup is low. If a deployment demand does fall on an active position with a designated IMA backup, mobilization of the IMA is likely to be on a voluntary basis. The combination of a position-specific focus and primary reliance on volunteerism suggests a very low probability that the program will be used as intended. The program would be more effective if it developed pools of IMAs available for backup in critical functions rather than individual IMAs for backup in specific positions.

Utility for Contingency Operations

As discussed above, demand for the deployment of ACS resources during the past decade has not been especially intense. Taking the IMA single/central manager programs as examples, Table

2.5 indicates that the medical, legal, and chaplain functions saw an average of less than 6 percent of AC personnel deployed at any one time and less than 2 percent of RC personnel deployed at any time. In the special investigations function, the deployment demand for AC personnel, at 10.5 percent, was somewhat higher than for other ACS functional areas, but even in this function, the RC deployment rate was under 2 percent.

Figures 2.7 and 2.8 examine how the distribution of IMA billets at the end of FY 2012 compared to the distribution of deployment demands experienced from FY 2002 to FY 2011 (the same data used in constructing Table 2.5). In Figure 2.7, we see that that more than half of the IMA billets are for officers, whereas only 17 percent of deployed days were experienced by officers. Although policies regarding IMA billets do not explicitly call for an officer-intensive distribution, this perhaps reflects a sense that the augmentation of officer billets generally has higher priority than the augmentation of enlisted billets.

Figure 2.7. Officer/Enlisted Distribution of IMA Billets (September 2012) and Deployment Demands (FY 2002–2011)

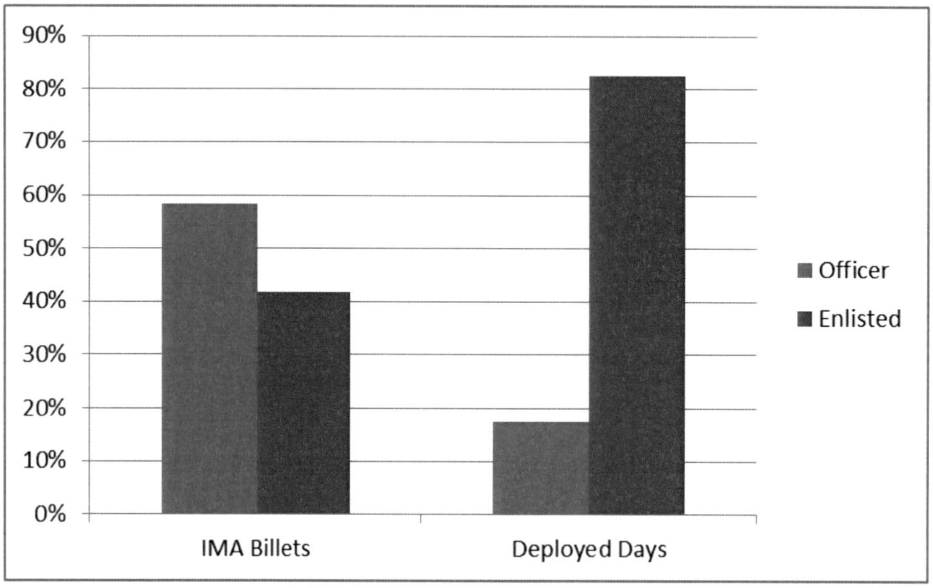

SOURCES: IMA billets: MPES; deployed days: DMDC and Air Force (multiple files), FYs 2002–2011.

In Figure 2.8, we array a subset of officer career fields (those with more than five IMA billets) from left to right in order of the average proportion of the AC population in the career field deployed at any one time. The figure also shows the proportion of total AC deployed days experienced by the AC population in the career field and the proportion of IMA billets distributed to the career field. Spikes in the proportion of IMA billets in a career field tend to coincide with the central and single manager programs. Overall, the distribution of IMA billets appears to be poorly matched to deployment demands.

Figure 2.8. Distribution of IMA Billets (September 2012) and Deployment Demands (FY 2002–2011), by Officer Career Field

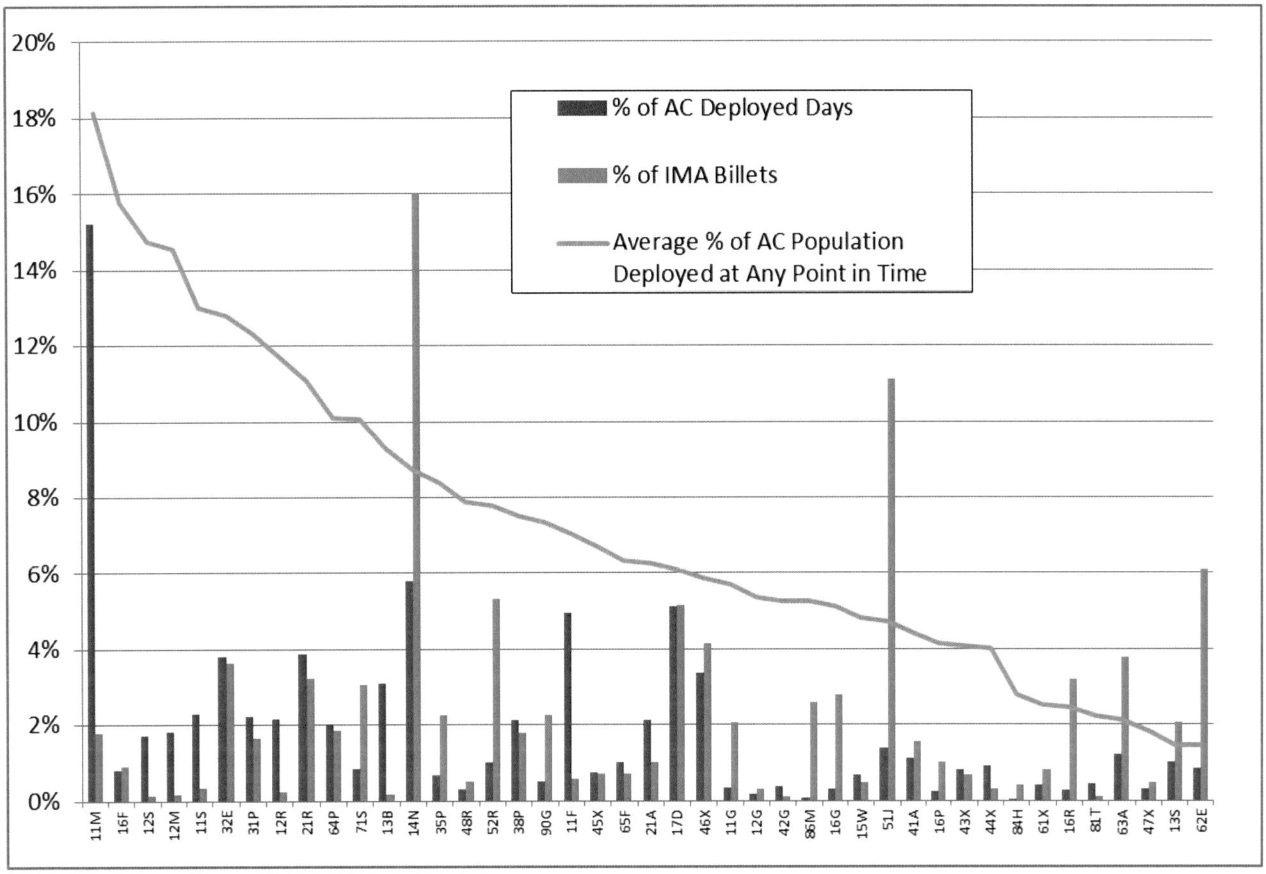

SOURCES: IMA billets: MPES, deployed days: DMDC and Air Force (multiple files), FYs 2002–2011.
NOTES: Career fields with five or fewer IMA billets are not included. Engineer and some medical career fields are collapsed to the two-digit level.

We also note that IMA authorizations are not taken into account in deliberate planning for wartime requirements.[17] Deliberate planning is based on UTC requirements, but IMA authorizations are not mapped to UTCs. Concentrating IMAs in larger pools rather than linking them to specific positions, as discussed above, might also enable the posturing of IMAs in UTCs, allowing a direct linkage to wartime requirements.

Utility for Other Purposes

Access to IMAs for specialized, technical, or scientific expertise or for economic reasons could benefit active units. However, the requirement for IMAs to drill for 24 to 48 periods per

[17] Joint campaign planning may be characterized as either *deliberate planning* or *crisis action planning*. Deliberate planning typically occurs during peacetime and is the process used when time permits the total participation of the Joint Planning and Execution Community. When the time available for planning is short and the near-term result is expected to be the actual deployment and/or employment of military forces, the planner uses crisis action planning procedures (Joint Forces Staff College, p. 4-3).

year and to avoid anything other than incidental contributions to active missions while drilling makes it unlikely that they can be used economically for occasional access to special expertise.

Summary

This chapter described Air Force manpower requirements processes and characterized their current state. It then highlighted aspects of these processes that appear to be problematic. We identified the following issues:

- The manpower requirements workforce may be insufficient in number or skill mix.
- At installations with both AC and RC representation, organizational and legislative restrictions block the synergistic merging of common functional activities and hence the development of integrated manpower standards.
- There are weak links in determining total force wartime ACS requirements, resulting in wartime UTC resources that are uneven across functions and, considering sparse usage during the past decade, likely overstated in the aggregate.
- The linkage between IMA authorizations and wartime requirements is tenuous.

In Chapter Four, we provide recommendations to address these issues, as well as the issues regarding qualitative requirements processes discussed in Chapter Three.

3. Qualitative Requirements Processes

Qualitative standards for military occupations are provided through the Air Force personnel classification system and, at a more detailed level, through training specifications. For civilian occupations, the federal civil service classification system serves the same purpose. MAJCOMs and other comparable organizations link these qualitative occupational standards to specific positions by associating AFSCs or civilian occupational codes with each position in the MPES file. Optionally, MPES may also identify required advanced education, language skills, special experience, security clearances, and other similar characteristics.

Military Requirements

While Air Force policy regarding military classification and training systems is determined by the Air Force Deputy Chief of Staff for Manpower, Personnel, and Services (AF/A1), career field managers provide the required technical expertise (AFI 36-2101, pp. 6–9; AFI 36-2201, p. 34). In addition to developing specialty descriptions, prerequisites, and qualifications, career field managers develop career field education and training plans (CFETPs) and Air Force job qualification standards (AFJQSs). These materials are common across all components (active, ANG, AFR) and are typically developed by active duty career field managers with coordination from reserve component counterparts.

The Air Force Classification System

The most comprehensive qualitative requirements for Air Force military personnel are conveyed through the Air Force military classification system. Air Force policy regarding classification rests on the concept of functional grouping: positions are grouped on the basis of similarities in "knowledge, education, training, experience, ability, and other common criteria" (AFI 36-2101, p. 5). The closely related concept of practical specialization recognizes that "[w]hile no one person is likely to perform all functions of an AFS at any one time, individuals can be developed to perform all duties and responsibilities of the various duty positions within an [Air Force specialty] at different times throughout a career" (AFI 36-2101, p. 5).

The system is structured using hierarchically arranged AFSCs consisting of four (officer) or five (enlisted) characters, with an optional prefix or suffix. In the officer AFSC, the digits represent, in order, a career group, utilization field, functional area, and qualification level. In the enlisted AFSC, the digits represent a career group, career field, career field subdivision, skill level, and specific AFSC. In both the officer and enlisted AFSCs, the prefix represents an ability, skill, special qualification, or system designator not restricted to a single AFSC, while the suffix (also called a *shredout*) represents particular equipment or functions within an AFSC.

AFSCs are used to identify both persons and positions. They provide the common link that allows current or projected personnel counts to be compared to current or projected manpower requirements for setting targets for recruiting, training, promotion, and other similar personnel programs and for distributing available personnel to organizations and geographical locations via Air Force assignment processes.

For each AFSC, a two-page description is provided in the Air Force Officer Classification Directory (AFOCD) or the Air Force Enlisted Classification Directory (AFECD), available in portable document format (PDF) on the Air Force myPers website.[1] Each description provides a specialty summary, duties and responsibilities, specialty qualifications (knowledge, education, training experience, etc.), and specialty shredouts. Qualifications may be mandatory or desirable.

Other elements of the Air Force classification system include special duty identifiers (SDIs), reporting identifiers (RIs), and special experience identifiers (SEIs). SDIs and RIs, identified by AFSC-like codes, are used to designate people and positions that do not clearly fall within a specific career field or fall within a condition rather than a duty (e.g., student, patient, or prisoner). Special experience identifiers are three-digit alphanumeric codes that identify useful experiences or training. While SEIs are not restricted to use with specific AFSCs, they are strongly oriented to specific functional areas and therefore may be practically applicable only to a limited range of AFSCs. Like AFSCs, SEIs are used to identify both people and positions. In the case of positions, applying an SEI is somewhat ambiguous—it may indicate either that the position imparts the designated experience or that it requires the designated experience (AFOCD, p. 250).

MPES-Based Requirements

Other qualifications required for a position may be specified in MPES position data. MPES has fields established to document the following requirements, although some fields have not been populated:

- security access
- grade
- advanced academic degree
- acquisition position type, category, career level, and special assignment
- financial management certification
- language
- regional expertise
- cultural capability
- personnel reliability program.

---

[1] The Air Force Personnel Center provides a variety of personnel services via the myPers website. Links to the AFOCD and AFECD can be found at https://gum-crm.csd.disa.mil/app/answers/detail/a_id/7759/kw/afocd and https://gum-crm.csd.disa.mil/app/answers/detail/a_id/7504/kw/afecd, respectively. Access to myPers is limited to personnel served by the site.

Training Plans and Job Standards

Career field education and training plans (CFETPs) provide a detailed specification of the job tasks encountered in a career field, along with an indication of how and to what depth each task must be trained for award and skill-level upgrading within an AFSC. They are required for enlisted specialties and optional for officer specialties (AFI 36-2201, p. 35). If necessary, even more detailed Air Force job qualification standards (AFJQMs) can be issued as supplements to CFETPs. CFETPs and AFJQMs are accessible on the Air Force e-Publishing website.

Civilian Requirements

Air Force civilian positions and employees, including non-appropriated fund positions and employees, generally are classified using federal civil service classification systems. The federal Handbook of Occupational Groups and Families (OPM, 2009) contains separate series for white-collar (general schedule) and blue-collar (trade, craft, or labor) positions. The handbook contains a brief description of each occupational series. More detailed specialization and grading information is contained in a separate position-classification or job-grading standard for each series.

As with military positions, civilian positions in MPES can carry some job qualification requirements. Additionally, civilian career fields may be included in corresponding officer or enlisted CFETPs.

Issues with Qualitative Requirements

While Air Force qualitative requirements processes seem sound in most respects, ongoing research within Project AIR FORCE has surfaced two issues, both relating to education requirements.

Conventionally, the AFOCD has listed either mandatory or desirable education prerequisites for various officer specialties. While this provided a sufficient basis for screening the eligibility of individual officers for entry into the specialty, it provided an inadequate basis for determining the desired distribution of academic disciplines in each entering cohort of officers. Other ongoing research has developed tables with greater specificity and prioritization of academic disciplines for each specialty, along with a recommended approach for determining a desired aggregate distribution.

MPES specifications for advanced academic degrees for selected positions have also come under scrutiny by another research effort in Project AIR FORCE. This research found a misalignment between the grades at which advanced academic degree requirements are established and the career points at which officers typically obtain their degrees. In general, the requirements tend to be linked to a lower distribution of grades than the available degree-holding population.

4. Practices in Selected Functional Areas

FAMs play pivotal roles in many of the processes described in Chapters Two and Three.[1] The FAM is the principal advisor to commanders and chiefs of major staff functions on the management and oversight of all personnel and equipment within a specific functional area to support operational planning and execution. They and their staffs precipitate process improvements and policy changes that can profoundly affect manpower standards. They, along with career field managers in the same function, determine the crew ratios and crew compositions that drive aircrew requirements and the scenarios that drive LCOM-derived aircraft maintenance requirements. They establish UTCs and should play a role in contingency and crisis action planning processes to determine if RC units are needed to meet contingency planning or rotational requirements. They track deploy-to-dwell rates for their functional areas and ensure the functional areas can meet ongoing operations within AEF rotations. They and their related career field managers are largely responsible for the contents of specialty descriptions in the AFOCD and AFECD. At the MAJCOM level, they strongly influence MPES-based qualitative specifications.

FAMs are enmeshed in these roles because they call for specialized functional expertise and judgment. Providing this functional depth in personnel requirements processes, however, also introduces the potential for inconsistency across functions. We found this to be particularly true in managing wartime requirements, where the oversight or review of functional manager actions by an enterprise-level office of responsibility was less evident than in other processes, such as establishing manpower authorizations or AFSC prerequisites. In addition to issues with inconsistency, we found little interaction among FAMs in their wartime planning. So, for example, changes to the size and phasing of UTCs in one functional area might not appropriately affect planning for other functional areas such as security, personnel, and services.

To further explore these issues, we contacted career field managers from three representative functional communities—civil engineering (CE), security forces, and manpower/personnel/services—to obtain descriptions of their involvement in personnel requirements processes.

Civil Engineering

The Air Force Civil Engineering community recognized, during the mid-1990s, that it and other ACS communities needed a more deliberate process to determine their total force wartime requirements (Cummings, 1997). The overall CE workforce process is more complex than for most ACS communities, involving multiple specialties in traditional CE, fire emergency service

[1] AF/A5XW is designated the Air Force Office of FAM Oversight (OFAMO) and AF/A1PR is designated as the OFAMO-Personnel (OFAMO-P) (AFI 10-401, p. 294).

(FES), emergency ordnance disposal, emergency management, and Rapid Engineers Deployable Heavy Operational Repair Squadron Engineer (RED HORSE) requirements. Therefore, developing wartime requirements is more complex. To size the community's requirements, the Air Force Civil Engineering Support Agency developed what became known as its *Blue Suit Review* process, supported by a model called the *Blue Suit Requirements Tool.*

The process uses the most stringent DoD-approved force structure planning scenario—currently, Integrated Security Construct–Scenario B (ISC-B)—combined with requirements for homeland defense, en route, headquarters staff, training, and civil engineers working outside the Air Force.[2] To arrive at some of these requirements, the community's planners examine Defense Planning Guidance (DPG) and other similar sources to assess wartime needs. The process also adds a factor to keep various specialties sustainable within career progression group guidelines. A key planning assumption is to provide for enough active duty forces to cover the first conventional campaign surge plus en route and active duty foundational requirements. The tool then allocates requirements by component (active duty, ANG, AFR), by MAJCOM, by base, by Air Force specialty, and by skill level. In the allocation process, the AC/RC mix in all specialties is fixed at 55 percent in the AC and dwell ratios are maintained at 1:2 and 1:5 for the AC and RC, respectively. The model does not consider garrison support requirements, which generally are met in part by military workforces needed for wartime requirements, with civilian and/or contractor manpower meeting the balance of needs.

This community has examined the AFPC/MA SMARTForce model and found that, at its present stage of development, it does not fully represent all wartime CE requirements. The community's Blue Suit Review process seems more comprehensive, but it may incorporate some assumptions that should be determined corporately by the Air Force rather than by each functional manager independently, such as planned force beddowns, career field sustainment factors, and the use of the AC only in the initial surge scenario (Air Force Civil Engineering Center, 2013). Also, its assumption of a fixed AC/RC mix unfortunately renders it unusable for the purpose of finding an AC/RC mix that better meets but does not exceed requirements. The strength of the SMARTForce model, when fully developed, will be the consistent treatment of these issues across all functions.

The CE FAM emphasizes that qualitative skills may need to be a factor in allocating manpower positions to the AC versus the RC. For example, it may be difficult for a TR to obtain or maintain the skills necessary for FES duty given their relatively limited training opportunities. Therefore, CE might be able to justify the added cost of FES AC manpower requirements in excess of garrison requirements. Again, a fixed ratio of AC to RC does not allow for these readiness considerations. Another potential barrier to determining the appropriate AC/RC mix is the inconsistent use of UTCs. In past contingency management actions, the RC CE community

---

[2] For this analysis, the force structure beddown required in ISC-B is reportedly based on output from AF/A9's TFE model.

has used tailored UTCs—providing less than the full manpower complement of a standard UTC. This may make it difficult for the AC to pass taskings on to the RC and measure RC capacity.

Security Forces

The security forces community—one of the larger functional communities in the Air Force—articulates its garrison and wartime requirements in the form of an inverted pyramid. As indicated in Figure 4.1, which depicts a slide used by the security forces community to brief its deployment requirements and resources, only about 10,000 out of 25,000 total AC authorizations are available to support deployment requirements, currently totaling about 2,400. Additionally, the 25,000 total authorizations do not include 6,700 unfunded positions needed to fully meet garrison requirements.

Figure 4.1. Security Forces Community View of Deployment Requirements and Resources

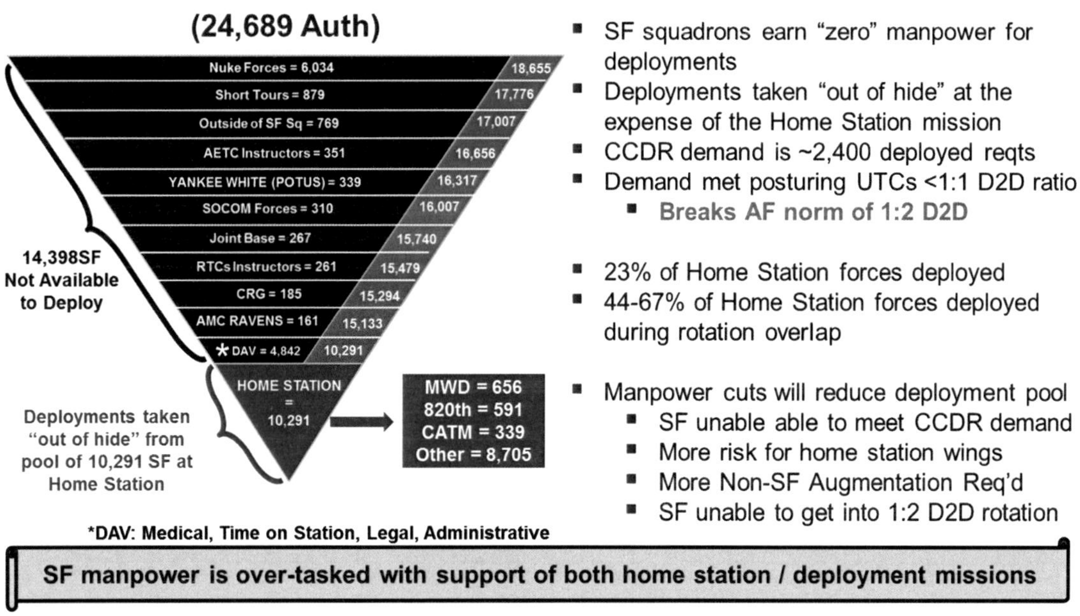

SOURCE: AF/A7SX.

NOTES: SF = security force; POTUS = president of the United States; SOCOM = United States Special Operations Command; DAV = deployment availability code; AETC = Air Education and Training Command; RTC = regional training center; CRG = contingency response group; AMC = Air Mobility Command; MWD = military working dog; CATM = combat arms training and maintenance; CCDR = combatant commander; D2D = deploy-to-dwell.

Security forces representatives indicated that about 350 of the 2,400 deployment requirements are met by RC personnel, drawing from about 10,000 RC authorizations. The AC deploy-to-dwell ratio thus appears to be 1:4 for individuals in the available pool.[3] However, at a

[3] If deployment demand is 2,400, of which 350 are met by the RC, the AC is continuously supplying about 2,000 deployed individuals from an available pool of 10,000, resulting in a 1:4 deploy-to-dwell ratio for those who are

UTC level, the ratio is reportedly less than 1:1. While the UTC-level ratio suggests that the community is overtasked, the individual-level ratio leads to a different conclusion—that the deployment stress does not exceed the 1:2 deploy-to-dwell ratio set by policy. At an individual level, the RC deploy-to-dwell ratio appears to be about 1:28, indicating a greater potential to relieve stress on the AC.

The security forces functional manager has not examined wartime needs to determine if the combination of AC and RC resources is sufficient. However, the wide variance between individual-level and UTC-level deploy-to-dwell ratios reported by this community suggests that much of the AC manpower in deployable units is unavailable for alignment with postured UTCs. In evaluating the adequacy of authorized manpower to meet deployment requirements at an acceptable deploy-to-dwell ratio, tools such as the SMARTForce model and the CE Blue Suit Requirements Tool must account for the conditions that limit the availability of individuals for deployment. We believe, however, that a more straightforward assessment of expected deploy-to-dwell ratios can be made by comparing required UTCs to postured UTCs.

Manpower, Personnel, and Services

The Air Force manpower, personnel, and services community is currently examining the adequacy of its resources to meet wartime needs. As with the CE community, this community considers DPG and other similar sources to assess its wartime requirements. Also, consistent with the CE process, it uses a force structure beddown obtained from AF/A9's TFE model. However, this community's methods of determining requirements are not as mature as CE's Blue Suit Review. The personnel FAM is just beginning to develop processes to size the community's total force wartime requirements, including determining how to phase UTCs to support deployment locations. Like the CE FAM, the personnel FAM said that his community is reviewing DPG and ISC-B, and developing planning factors and assumptions itself rather than obtaining them from some source providing information to all FAMs.

The community finds that wartime deployment demand for its manpower and personnel subfunctions is relatively light, while demand for its services subfunction is relatively heavy. However, the community has not fully evaluated the need for RC manpower resources in light of these wartime demands. The FAM highlighted the role of the career field manager in determining and structuring garrison manpower requirements as having a significant impact on the AC forces available for deployment. The service function, for example, in both garrison and wartime requirements, is highly dependent on planning for the role of contractors (e.g., will local contractors provide food service at deployed locations after the predetermined duration of the operation?). The FAM repeatedly raised concerns about the lack of an Air Force process for determining overall wartime requirements in a function, instead relying on an inventory-based

---

continually in the available pool. If the approximately 5,000 personnel temporarily unavailable for deployment, as indicated by DAVs, are included in the computation, the individual-level AC ratio is 1:6.5.

process that postures all manpower authorizations without regard for whether they are required for wartime purposes. He advocates a return to a more deliberate planning process—a process like FORSIZE/BLA or TFA—last accomplished in 2001.

The personnel FAM also expressed concerns about the lack of guidance for determining "minimum critical home station requirements" (i.e., those positions coded as being required at the home station during wartime and therefore not available for deployments). This designation is inconsistently applied across MAJCOMs and even bases.

5. Summary and Recommendations

In this chapter, we summarize the findings discussed in previous chapters and offer our recommendations for improving personnel requirements processes. Our recommendations include

- improvements in existing requirements processes
- refinements in wartime requirements policies and practices
- steps to provide greater flexibility in using RC resources
- ideal and practical approaches to integrating requirements across workforces
- closer examination of IMA requirements
- improved specification of officer education requirements.

Improving Existing Manpower Requirements Processes

To be fully equal to its purpose in strategic human capital management, the Air Force manpower requirements process requires some refinements. The proportion of total manpower requirements covered by standard processes is much less than 100 percent. Many available standards are over a decade old, and there is decreasing confidence that they still reflect the needs of changing missions, technologies, and organizational concepts. Both of these gaps seem to be related to an insufficient quantity and/or quality of resources devoted to the development of manpower standards and other requirements processes. Another possibility lies in redesigning manpower requirements development processes, perhaps including a prudent trade-off of measurement precision.

Refined Policies and Practices for Determining Wartime Requirements

The interplay of garrison and deployment requirements is managed primarily by FAMs with insufficient corporate oversight and guidance. No Air Staff focal point has been identified as a single source for providing functional managers with a common and sufficiently detailed set of planning assumptions regarding deployment requirements. As a result, each functional manager develops an independent sense of deployment demands, resulting in an unbalanced assessment of those demands. Additionally, while TR requirements provide the primary means of balancing resources with deployment requirements, there are no institutionalized processes by which AC and RC functional managers or AC and RC manpower managers can interface.

We found wartime planning, and hence determining the required size of RC manpower, to be in a relatively poor state in ACS functions. For operational capabilities, the AC/RC mix is explicitly considered in the planning and programming actions of the Air Force Corporate Structure with considerable supporting analysis from AF/A9, which is largely developed using its TFE model. However, the TFE model does not provide programming-quality ACS wartime requirements outputs by function at either the individual manpower authorization or UTC level.

While it is possible that the lack of rigor in determining ACS wartime requirements results in an undersupply of resources in some cases, the body of evidence we reviewed (see Table 2.5 and accompanying discussion) suggests broad-based oversupply. However, the sizing of wartime requirements seems to rest on several questionable assumptions, which we believe warrant policy changes or clarifications.

First, we note that RC wings generally mirror the structure of active wings, even though their mission is usually on a much smaller scale than active wings. A typical AC fighter wing, for example, consists of three squadrons of 24 aircraft each, while an RC fighter wing typically has one squadron of 18 aircraft. The RC wing often has the same array of functional squadrons as an AC wing, in addition to its wing and group command elements. Most of this leadership structure is not needed for wartime deployments, i.e., it does not correspond to UTC requirements unmet by the AC. Peacetime organizing, training, and equipping could likely also be done effectively with a much flatter, leaner organizational structure.

Second, cutting in the direction of a possible greater need for RC deployable resources, we question the reasonableness of a 1:2 deploy-to-dwell ratio for the AC and a 1:5 mobilization-to-dwell ratio for the RC. While these rates might be tolerable for relatively brief periods, we suspect that the Air Force would be seriously strained by sustained deployment and mobilization at those rates, particularly with respect to garrison support within the AC. The level of strain would also depend on how these rates are measured. If a 1:2 deploy-to-dwell ratio means that one-third of every garrison's deployable manpower is continually deployable, we believe the effect during a prolonged scenario requiring deployment would be a serious degradation of garrison support services and continuously overworked garrison support workforces. If this ratio means that one or more UTCs is deployed from each squadron during every third AEF rotation, the effects would be much less severe. For planning purposes, the definitions of the ratios need to be sharpened. Depending on the definitions adopted, the ratios may need adjustment to be considered sustainable.

Third, the process of posturing UTCs needs additional policy guidance. The Air Force's policy of posturing all available resources in UTCs tends to overstate UTC availability because it fails to account for the need for continued garrison support requirements during wartime. UTC posturing guidance should explicitly state that the minimum manpower needed to support the garrison over a sustained period should remain unpostured.

Greater Flexibility in Using RC Resources

We have argued that the legal requirements for RPA-funded activity to be focused on the training and administration of reserve forces results in more training than is needed in ACS functions. RC resources would be more useful if RPA-funded activity could be utilized without legal restrictions to support active missions.

An Ideal Approach for Integrating ACS Requirements Across Workforces

The underlying objective of the Air Force manpower system is efficiency—"the effective and economical use of manpower" (AFI 38-201, p. 7). In the spirit of this principle, and recognizing that a workload often can be accomplished by any of several alternative workforces (AC, RC, civilian, or contractor) that vary in cost, we posit that cost minimizing is an ideal approach to integrating the requirements for these various workforces. We further posit, based on what we heard from functional managers, that the integration process should consider garrison, wartime, and career field sustainment needs. The task, given that these needs are known, is to find the least-cost workforce mix that meets them.

A general approach to integrating ACS requirements across workforces, consistent with broad Air Force policies, is as follows:

- Distribute operating force missions across components and garrisons so as to minimize cost while meeting other important constraints.[1]
- Size AC ACS garrison demands, which can be met by any combination of military, civilian, and contractor workforces.
- Size total force wartime ACS demands, which can be met by AC military and RC traditional-reservist workforces.
- Meet as much of the wartime ACS demand as possible using AC military forces deployed or employed from garrisons.
- Size RC garrison ACS resources to meet remaining wartime requirements.
- Use civilian or contractor workforces for any ACS garrison requirements not needed for wartime requirements or other military essentiality considerations.

As a starting point for a full-featured process, we will assume that workload demands can be expressed in terms of military grades and specialties derived from UICs (garrison demands) and UTCs (wartime demands).[2] We also stipulate that this approach will not be used in place of conventional manpower standards or other requirements processes, but it will be used to assess the relative efficiency of existing workforce mixes and influence manpower programming and execution decisions that affect the mix. Thus, we can ignore minor differences in man-hour

---

[1] Rational distribution of operating forces across components and garrisons is itself a major challenge; addressing it is beyond the scope of this document.

[2] Measuring workloads in this manner can only be approximated. In MPES, all civilian and contractor manpower equivalent (CME) authorizations carry an equivalent military AFSC. Grade-range approximations can be derived from the skill level of the AFSC.

availability factors among various full-time workforces.[3] Additionally, since the decisions that the approach will influence are of primary concern to functional managers, we outline an approach that is intended to be applied to requirements within a functional community.

The approach we recommend takes the form of an optimization model that varies the count of authorizations in each grade, AFSC, and workforce type (active, reserve full-time, reserve part-time, civilian, and CME) so as to minimize cost while meeting various requirements-based constraints. The model requires, as inputs, the composite cost of each grade/AFSC/workforce combination. On the requirements side, inputs would include counts of UTCs needed to meet surge and rotational deployment demands, along with the grade and AFSC composition of each UTC and sums of garrison UICs by grade, AFSC, continental United States (CONUS)/overseas location, and deployable/nondeployable status. For career field sustainment purposes, the model would also require the appropriate ratios between grades and the feeder/capper relationships of converging specialty ladders in the functional area. For the purpose of estimating UTC capacities, historical or objective deployment availability rates and UTC-level deploy-to-dwell target rates would also be required. Additionally, to avoid degrading garrison mission performance to too great a degree, inputs would include maximum mission degradation factors (i.e., the maximum proportion of an AFSC that should be deployed from a location at any time).

In general, the model would meet UTC demands using either the AC or RC, creating as many traditional reservist positions as needed. Since civilian authorizations are generally less expensive than military authorizations, it would use only as many military authorizations as needed to fill out UTC requirements, plus a factor to account for deployment nonavailability, and any increases in specific grades needed to maintain career field sustainment. Remaining garrison requirements would be filled by civilian or contractor workforces, whichever were cheaper.

For the AC portion of the workforce, the model would also require a constraint to insure that military workforces in CONUS garrison units are sufficient to support the rotation to overseas garrison units.

In this model, RC requirements would be treated as a variable rather than an input. The TR requirement would vary to support UTC requirements that exceeded the feasible level of UTC posturing from AC garrison workforces. The RC full-time workforce requirement would be derived partially from garrison support demands (an input to the model) and partially from the size of the TR requirement generated by the model, given the full-time RC workforces' responsibilities to train and administer the part-time RC workforces.

---

[3] Monthly man-hours available to primary duty for personnel on normal 40-hour workweek schedules are 150.7 hours for military personnel, 143.3 hours for CONUS-based civilians, and 149.9 hours for overseas-based civilians (AFI 38-201, Table A3.1).

A Practical Approach for Integrating ACS Personnel Requirements

A model at the level of complexity described above would be challenging to construct, maintain, and routinely exercise. A more practical approach, but one less likely to reveal the most efficient workforce mix, would employ a series of partial assessments and adjustments of the workforce mix, including

- determining, in cooperation with OSD Cost Assessment and Program Evaluation (OSD CAPE), required forces and beddowns for designated wartime scenarios. Ideally, this would be accomplished through systematic updates of OPLANs and TPFDDs.
- using a distribution of operating missions by component and garrison that emerges from various planning, programming, and political processes
- continued use of standard manpower requirements and Corporate Structure processes to assess the need for and program AC garrison ACS manpower requirements
- a UTC demand summary (a tool that does not currently exist) listing the maximum required counts of UTCs, by type, in various phases of designated scenarios. For each phase, the summary should also indicate whether the mobilization of reserve resources is assumed to be authorized and whether the phase is assumed to be in a surge (non-rotational) or post-surge (rotational) mode.[4]
- continued development and posturing of UTCs to meet wartime needs
- continued use of AF/A9's TFE model to estimate force structure and deployment demands for operational and maintenance units, but with the aforementioned UTC demand summary as an input
- a fully developed SMARTForce model to estimate deployment and home station warfighting demands for ACS forces, with the UTC demand summary as an input
- a tool for easily comparing postured UTCs in the War and Mobilization Planning, Volume 3 (WMP-3), Part 2 UTC Availability database (see AFI 10-401, pp. 77–78) to requirements in the UTC demand summary for any designated scenario
- a distribution of the number of UTCs to be supplied by each AC MAJCOM (based on UTC availability reflected in WMP-3) for each type of ACS UTC in the UTC demand summary, with the remainder assigned to the RC (broken out by ANG and AFR)
- a periodic comparison of postured RC ACS UTCs to RC assignments in the UTC demand summary, accompanied by an adjustment of TR authorizations to match the supply of RC UTCs to the demand
- periodic checks to insure that CONUS/overseas rotation among garrison units is maintained at an acceptable level
- periodic alignment of grades within each career field using prescribed career progression group procedures (AFI 38-201, pp. 30–34).
- periodic military-to-civilian or civilian-to-military conversions to better accommodate UTC posturing, CONUS/oversea rotations, and grade progressions

---

[4] We envision an iterative process for determining this demand with input from each functional area and across functional areas, since ACS functions are interdependent, e.g., the number of personnel in a given function needed at a location is dependent on how many personnel the other functions require at that location.

- periodic contracting/in-house workforce cost comparison studies to determine if outsourcing is appropriate.

A concrete example might help to illustrate how this approach would be applied in determining the ACS manpower needed to supply required total force wartime capabilities. Assume that within a given function and for a UTC of a specific type, the maximum wartime requirements derived from approved scenarios are 50 instances of the UTC during surge phases and 40 instances during steady-state phases. Also assume that AC manpower is sufficient to posture 90 instances of the UTC. The surge requirement, accordingly, can be met with no RC activations. Even though fewer instances of the UTC are needed during steady-state phases in this example, rotational requirements drive up the need for many more postured UTCs than would be required in a surge. Thus, at a 1:2 deploy-to-dwell ratio, the 90 available AC UTCs could provide 30 of the 40 UTCs required in the steady-state phase. The remaining 10 steady-state requirements would be met from the RC, which, at a 1:5 activation-to-dwell ratio, would need sufficient manpower to posture at least 60 instances of the UTC.[5] Since all of the available AC resources, and then some, are needed to meet wartime steady-state requirements, any civilian authorizations or contractor-supplied manpower in the function should be examined to see if conversions to military manpower would allow for additional instances of the UTC to be postured. If so, that would reduce the need for RC-supplied instances of the UTC, allowing for a reduction in RC manpower and likely some savings in overall manpower costs in the affected functional area.

Organizational Issues

To implement the approach outlined above, it would be necessary to identify responsibility for key actions in this series of assessments and adjustments. As the office of primary responsibility for contingency planning (AFI 38-401, pp. 49–50), AF/A5XW should play a critical role, particularly in the construction of the UTC demand summary proposed here. AF/A1M and its counterparts in the RC would also play a role in assessing the adequacy of UICs to meet UTC demands. Functional managers would, of course, continue to play a role in designing UTCs and incorporating them in TPFDDs, force modules, and other wartime planning constructs. MAJCOMs and units would play a key role in posturing UTCs. AF/A9 would continue its analytic support role. Finally, the Air Force Corporate Structure, for which the Air Force Directorate of Programs (AF/A8P) bears much of the administrative responsibility, would use UTC supply-and-demand analyses as a key input into its resourcing of the RC.

In discussions with AF/A5XW staff, we heard that construction of a UTC demand summary would present complexities that they are not equipped to address. Their attention is consumed by the process of managing the current supply of deployable resources to meet current demands.

---

[5] Generally, more than 60 UTCs would be required because pre-deployment training requirements reduce the time available for deployment during an RC UTC's period of activation.

Given the uncertainties of future combatant commander needs, they see no clear avenue toward the development of future UTC requirements, particularly in ACS functions. Their focus, in the parlance of wartime planning, is on crisis action planning rather than deliberate planning.

We recognize that requirements planning is difficult under conditions of uncertainty, but resourcing decisions can and should be made and informed by the best available risk assessments. Failure to articulate requirements results in uneven resourcing across functions. Importantly, we see evidence that many ACS functions are over-resourced relative to their wartime requirements, i.e., they can posture more UTCs than are needed at acceptable deploy-to-dwell and activation-to-dwell ratios. If these functions were more prudently resourced, the Air Force could retain greater force structure, and hence greater combat capability, in the face of impending budget cuts.

As of the time of this writing, the Air Staff is contemplating a reduction of 20 percent of its headquarters staff, consistent with guidance from the Secretary of Defense (Shoop, 2013). Streamlining the headquarters structure to accommodate this reduction could result in the redistribution of existing wartime planning responsibilities. Given the significant resource demands predicated on wartime requirements, we recommend that a key objective of the reorganization should be the revitalization of deliberate wartime planning capabilities.

Better Use of IMAs

We find that the policies governing the use of IMAs could be refined so as to shape a program with much greater potential to contribute to wartime or other Air Force needs. For the purposes of meeting wartime requirements, we believe that guidelines are needed to determine which active positions should be eligible for backfill or augmentation. In the spirit of Sneed and Kilmer (2012), and for clarity in justifying positions, we suggest differentiation of MSI codes to indicate the primary purpose of a position—back-fill, recurring deployment, augmentation, or other. For enhanced connectedness to wartime requirements, we recommend associating pools of IMAs with specific functions rather than individual IMAs with specific positions and, potentially, posturing IMA authorizations. The pools of IMAs should be concentrated in functions with wartime demands that are not easily met using UTC-postured AC manpower. Finally, we recommend using standard manpower allocation procedures to provide more voice to using commands in managing IMA authorizations.

Improved Specification of Officer Education Requirements

Among qualitative requirements, both undergraduate and graduate education specifications would benefit from closer attention. Undergraduate requirements contained in officer specialty descriptions need to be made more specific so as to provide a desired distribution of academic disciplines in officer accession cohorts. To promote better utilization of earned degrees, advanced academic degree requirements in MPES records need to be realigned to higher grades to better match the grade distribution of advanced academic degree holders.

Appendix A. Manpower Standard Implementation Codes

MSI Category	MSI Code	MSI Definition	Authorizations
Manpower standard	G	Manpower standard	64,908
	H	Air Force- approved legacy manpower standard	53,248
	J	MAJCOM-approved legacy manpower standard	21,021
Crew ratio	F	Crew ratio	16,084
LCOM	L	LCOM capability manpower standard	49,007
Tech estimate	X	Medical product line analysis transition team (Platt)-based technical estimate	35,418
	Y	Technical estimate	10,856
UTC	T	UTC requirement (ANG/Reserve use only)	123,034
Hq Staffs	D	Above wing headquarters staff and covered by manpower standard	892
	E	Above wing headquarters staff and sized by a service level	26,465
	N	Outside Air Force	27,447
IMA	V	Individual mobilization augmentee	8,674
Other	B	Foreign military sales (FMS)	2,929
	M	Competitive sourcing (MEO determined)	3,808
	K	Not covered by manpower standard but is agreed through memorandum of understanding or treaty	9,465
No standard process	0	No title – migration from previous system	4
	A	Non A76 contractor manpower equivalent requirement	155
	P	No title - migration from previous system	3
	Q	Not covered by mnpwr std but wrkld is AF directed	50,354
	R	Not covered by mnpwr std but wrkld is MAJCOM directed	57,832
	U	No title - migration from previous system	6
	W	Deployment participation - MAJCOM funded	43
	Z	Pending validation	93,640
		Total	655,293

SOURCE: Data extracted from MPES as of the end of fiscal year 2012.
NOTE: Includes funded authorizations only and excludes permanent party student authorizations.

Appendix B. Manpower Requirements Processes

The flow chart on the following pages depicts the major decision or administrative processes, data, and offices of primary responsibility involved in the development of various kinds of manpower requirements.

Although there are some temporal relationships in the various subprocesses, much of the overall process is cyclical and simultaneous. Decisions mutually affect one another over time, such that the overall system has at least the potential to evolve toward optimality.

Manpower Requirements Processes

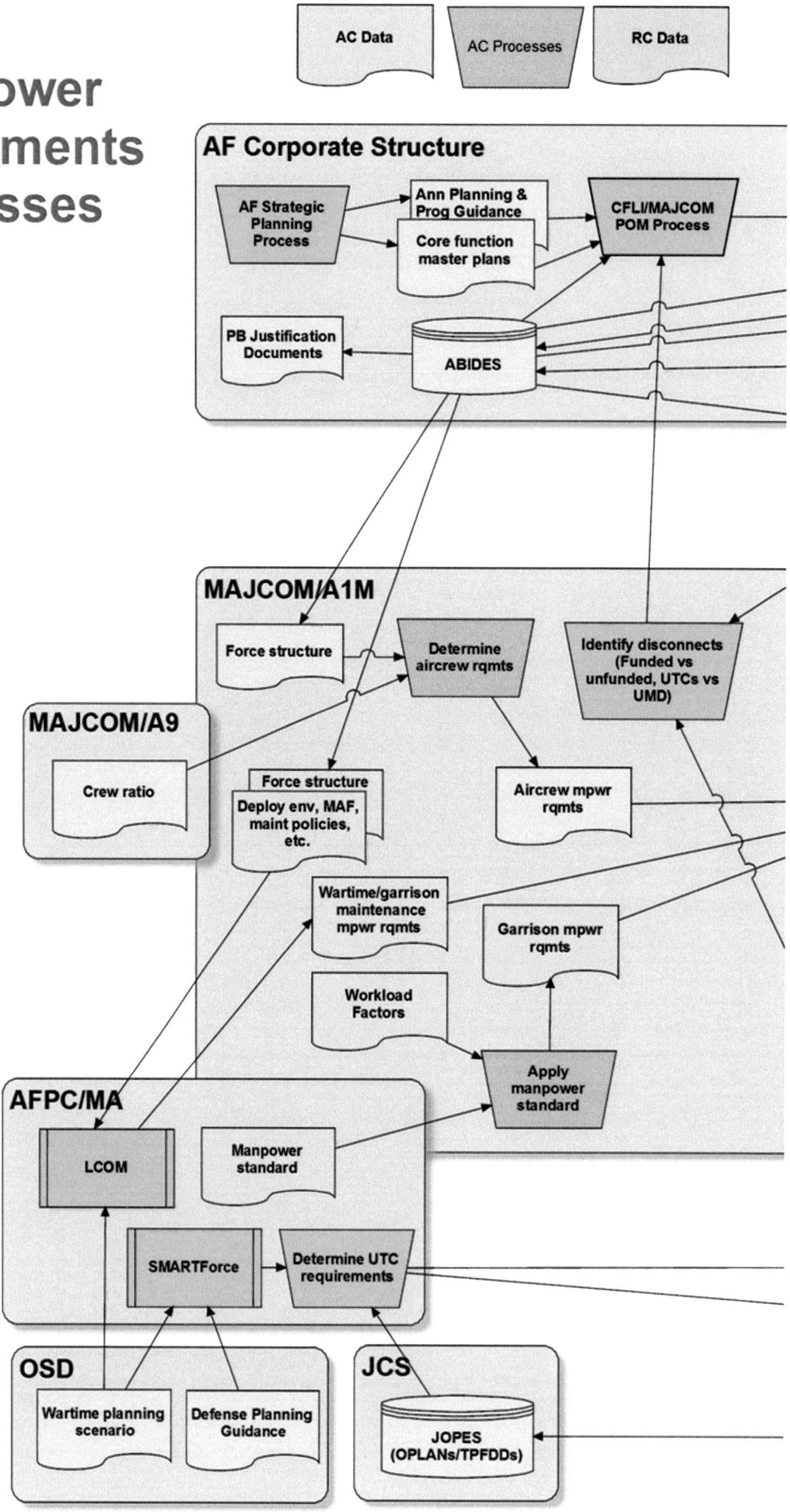

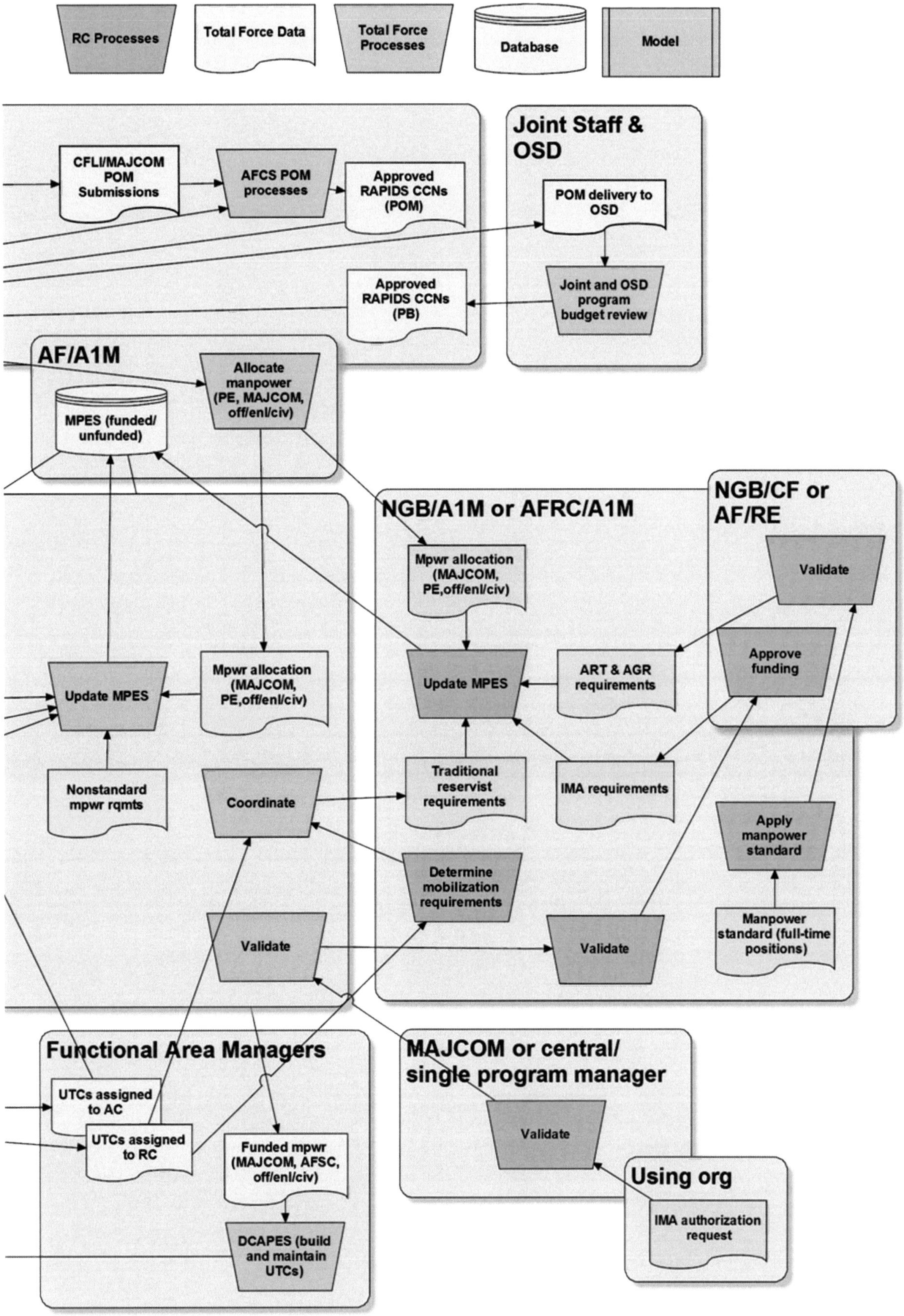
RC Processes
Total Force Data
Total Force Processes
Database
Model
Joint Staff & OSD
CFLI/MAJCOM POM Submissions
AFCS POM processes
Approved RAPIDS CCNs (POM)
POM delivery to OSD
Approved RAPIDS CCNs (PB)
Joint and OSD program budget review
AF/A1M
Allocate manpower (PE, MAJCOM, off/enl/civ)
MPES (funded/ unfunded)
NGB/A1M or AFRC/A1M
NGB/CF or AF/RE
Mpwr allocation (MAJCOM, PE,off/enl/civ)
Validate
Update MPES
Mpwr allocation (MAJCOM, PE,off/enl/civ)
Update MPES
ART & AGR requirements
Approve funding
Nonstandard mpwr rqmts
Coordinate
Traditional reservist requirements
IMA requirements
Apply manpower standard
Determine mobilization requirements
Validate
Validate
Manpower standard (full-time positions)
Functional Area Managers
UTCs assigned to AC
UTCs assigned to RC
Funded mpwr (MAJCOM, AFSC, off/enl/civ)
DCAPES (build and maintain UTCs)
MAJCOM or central/ single program manager
Validate
Using org
IMA authorization request

References

AFI—*See* Air Force Instruction.

AFMAN—*See* Air Force Manual.

AFPC/MA—*See* Air Force Military Personnel Center, Directorate of Manpower.

AFRCI—*See* Air Force Reserve Command Instruction.

Air Force Civil Engineering Center, "Civil Engineer Blue Suit Review Concept: Engineer Requirements Needed to Support Our Air Force," briefing, June 2013.

Air Force e-Publishing website, undated. As of November 19, 2013: http://www.e-publishing.af.mil

Air Force Instruction 10-402, *Mobilization Planning*, Washington, D.C., May 1, 2012. As of February 12, 2014: http://static.e-publishing.af.mil/production/1/af_a3_5/publication/afi10-402/afi10-402.pdf

Air Force Instruction 16-501, *Control and Documentation of Air Force Programs,* Washington, D.C., August 15, 2006. As of March 11, 2013: http://static.e-publishing.af.mil/production/1/af_a3_5/publication/afi16-501/afi16-501.pdf

Air Force Instruction 36-2101, *Classifying Military Personnel (Officer and Enlisted)*, Washington, D.C., June 25, 2013. As of June 25, 2013: http://static.e-publishing.af.mil/production/1/af_a1/publication/afi36-2101/afi36-2101.pdf

Air Force Instruction 36-2201, *Air Force Training Program*, Washington, D.C., September 15, 2010. As of June 25, 2013: http://static.e-publishing.af.mil/production/1/af_a1/publication/afi36-2201/afi36-2201.pdf

Air Force Instruction 38-201, *Management of Manpower Requirements and Authorizations*, Washington, D.C., May 14, 2013. As of August 7, 2013: http://static.e-publishing.af.mil/production/1/af_a1/publication/afi38-201/afi38-201.pdf

Air Force Instruction 38-204, *Programming USAF Manpower*, Washington, D.C., August 1, 1999. As of June 27, 2013: http://static.e-publishing.af.mil/production/1/af_a1/publication/afi38-204/afi38-204.pdf

Air Force Instruction 90-1001, *Responsibilities for Total Force Integration*, Washington, D.C., May 29, 2007. As of March 11, 2013: http://static.e-publishing.af.mil/production/1/af_a8/publication/afi90-1001/afi90-1001.pdf

Air Force Manual 38-201, Vol. 1, *Air Force Management Engineering Program (MEP)—Processes*, Washington, D.C., March 29, 2002. As of March 13, 2013: http://static.e-publishing.af.mil/production/1/af_a1/publication/afman38-208v1/afman38-208v1.pdf

Air Force Manual 38-201, Vol. 3, *Air Force Management Engineering Program (MEP)—Logistics Composite Model (LCOM)*, Washington, D.C., March 31, 1995. As of March 13, 2013:
http://static.e-publishing.af.mil/production/1/af_a1/publication/afman38-208v3/afman38-208v3.pdf

Air Force Military Personnel Center, Directorate of Manpower, "Strategic Manpower Assessment of Requirements—Total Force (SMART Force)," briefing, Randolph AFB, Tex., July 22, 2012.

Air Force Reserve Command Instruction 38-201, *Manpower Authorization, Allocations, and Change Requests*, Washington, D.C., June 23, 2000. As of March 13, 2013: http://static.e-publishing.af.mil/production/1/afrc/publication/afrci38-201/afrci38-201.pdf

Air National Guard Instruction 38-201, *Determining Manpower Requirements*, Washington, D.C., April 27, 2009. As of March 13, 2013: http://static.e-publishing.af.mil/production/1/ang/publication/angi38-201/angi38-201.pdf

ANGI—*See* Air National Guard Instruction.

Cummings, Gregory A., "Civil Engineer Blue Suit Wartime Requirements Review," *The CE Magazine*, Vol. 5, No. 3, Fall 1997.

Department of Defense Directive 1235.10, *Activation, Mobilization, and Demobilization of the Ready Reserve*, November 26, 2008. As of February 13, 2014: http://www.dtic.mil/whs/directives/corres/pdf/123510p.pdf.

Department of Defense Instruction 5100.73, *Major DoD Headquarters Activities*, December 1, 2007. As of March 13, 2013: http://www.dtic.mil/whs/directives/corres/pdf/510073p.pdf

DoDD—*See* Department of Defense Directive.

DoDI—*See* Department of Defense Instruction.

Goulka, Jeremiah, and Tom Englehardt, "Is Lockeed Martin Sequester-Proof?" antwar.com, March 11, 2013. As of March 11, 2013: http://original.antiwar.com/engelhardt/2013/03/10/is-lockheed-martin-sequester-proof/

Joint Forces Staff College, *The Joint Staff Officer's Guide 2000*, Washington, D.C.: National Defense University, Publication 1, 2000. As of February 13, 2014: http://www.uscg.mil/directives/cim/3000-3999/CIM_3020_15D.pdf

Mills, Patrick, John G. Drew, John A. Ausink, Daniel M. Romano, and Rachel Costello, *Balancing Agile Combat Support Manpower to Better Meet the Future Security Environment*, Santa Monica, Calif.: RAND Corporation, RR-337-AF, 2014. As of June 2014: http://www.rand.org/pubs/research_reports/RR337.html

Office of Management and Budget, "Performance of Commercial Activities," Circular No. A-76 Revised, May 29, 2003. As of November 19, 2013: http://www.whitehouse.gov/omb/circulars_a076_a76_incl_tech_correction

Public Law 112-10, Department of Defense and Full-Year Continuing Appropriations Act, 2011. As of November 19, 2013: http://www.gpo.gov/fdsys/search/pagedetails.action?browsePath=112%2FPUBLIC%2F[%3B99]&granuleId=&packageId=PLAW-112publ10&fromBrowse=true

Shoop, Tom, "Former Air Force Secretary to Reduce Hagel's Staff by 20 Percent," *Government Executive*, April 27, 2013. As of September 9, 2013: http://www.defenseone.com/management/2013/08/former-air-force-secretary-reduce-hagels-staff-20-percent/69479/

Secretary of Defense Memorandum, "Utilization of the Total Force," January 19, 2007. As of February 13, 2014: http://ra.defense.gov/documents/rtm/SECDEF%20Utilization%20of%20Total%20Force%2019Jan07.pdf

Snyder, Don, and Patrick Mills, *Supporting Air and Space Expeditionary Forces: A Methodology for Determining Air Force Deployment Requirements*, Santa Monica, Calif.: RAND Corporation, MG-176-AF, 2004. As of November 19, 2013: http://www.rand.org/pubs/monographs/MG176.html

United States Code, Title 10, Armed Forces, 2011. As of November 19, 2013: http://www.gpo.gov/fdsys/search/pagedetails.action?collectionCode=USCODE&searchPath=Title+10&oldPath=&isCollapsed=true&selectedYearFrom=2011&ycord=418&browsePath=Title+10&granuleId=&packageId=USCODE-2011-title10&collapse=true&fromBrowse=true

———, Title 32, National Guard, 2011. As of November 19, 2013: http://www.gpo.gov/fdsys/search/pagedetails.action?collectionCode=USCODE&searchPath=Title+10&oldPath=&isCollapsed=true&selectedYearFrom=2011&ycord=418&browsePath=Title+32&granuleId=&packageId=USCODE-2011-title32&collapse=true&fromBrowse=true